The Ecology of
Woodland Creation

The Ecology of
Woodland Creation

Edited by

RICHARD FERRIS-KAAN

Wildlife & Conservation Research Branch, The Forestry Authority, UK

JOHN WILEY & SONS
Chichester • New York • Brisbane • Toronto • Singapore

Published 1995 by John Wiley & Sons Ltd,
Baffins Lane, Chichester,
West Sussex PO19 1UD, England

Telephone National   01243 779777
International   (+44) 1243 779777

Other Wiley Editorial Offices

John Wiley & Sons, Inc., 605 Third Avenue,
New York, NY 10158–0012, USA

Jacaranda Wiley Ltd, 33 Park Road, Milton,
Queensland 4064, Australia

John Wiley & Sons (Canada) Ltd, 22 Worcester Road,
Rexdale, Ontario M9W 1L1, Canada

John Wiley & Sons (SEA) Pte Ltd, 37 Jalan Pemimpin #05-04,
Block B, Union Industrial Building, Singapore 2057

Library of Congress Cataloging-in-Publication Data

The ecology of woodland creation / edited by Richard Ferris-Kaan.
    p.   cm.
    Modified versions of papers presented at the British Ecological Society's Forest
Ecology Group symposium in Leicester, Apr. 1993; with additional material.
    Includes bibliographical references and index.
    ISBN 0-471-95484-5
    1. Forest ecology — Congresses.   2. Restoration ecology — Congresses.
    3. Forest landscape design — Congresses.   4. Afforestation — Congresses.
    5. Reforestation — Congresses.   6. Forest ecology — Great Britain —
Congresses.   7. Restoration ecology — Great Britain — Congresses.   8. Forest
landscape design — Great Britain — Congresses.   9. Afforestation — Great
Britain — Congresses.   10. Reforestation — Great Britain — Congresses.
I. Ferris-Kaan, Richard.
II. British Ecological Society. Forest Ecology Group.
QH541.5.F6E28 1995   94-35539
574.5'2642—dc20   CIP

British Library Cataloguing in Publication Data

A catalogue record for this book is available from the British Library

ISBN 0-471-95484-5

Typeset in 10/12pt Times from editor's disk by Saxon Graphics Ltd, Derby
Printed and bound in Great Britain by Bookcraft Ltd, Bath, Avon

Contents

List of Contributors

S. Bell   The Forestry Authority, 231 Corstorphine Road, Edinburgh, EH12 7AT

G.P. Buckley   Dept. of Horticulture, Wye College (University of London), Wye, Ashford, Kent TN25 5AH

E.V.J. Cohn   School of Applied Sciences, University of Wolverhampton, Wulfruna Street, Wolverhampton WV1 1SB

J.R. Flowerdew   Department of Zoology, University of Cambridge, Downing Street, Cambridge CB2 3EJ

R.J. Fuller   British Trust for Ornithology, The Nunnery, Thetford, Norfolk IP24 2PU

R.M.A. Gill   Wildlife and Conservation Research Branch, The Forestry Authority, Alice Holt Lodge, Wrecclesham, Farnham, Surrey GU10 4LH

S. Gough   British Trust for Ornithology, The Nunnery, Thetford, Norfolk IP24 2PU

J. Gurnell   School of Biological Sciences, Queen Mary and Westfield College, University of London, Mile End Road, London E1 4NS

R. Harmer   Silviculture (South) Branch, The Forestry Authority, Alice Holt Lodge, Wrecclesham, Farnham, Surrey GU10 4LH

J.A. Harris   Environment and Industry Research Unit, Department of Environmental Sciences, University of East London, Romford Road, London E75 4LZ

T.C.J. Hill   Environment and Industry Research Unit, Department of Environmental Sciences, University of East London, Romford Road, London E75 4LZ

G. Kerr   Silviculture (South) Branch, The Forestry Authority, Alice Holt Lodge, Wrecclesham, Farnham, Surrey GU10 4LH

R.S. Key   English Nature, Northminster House, Peterborough PE1 1UA

J.H. Marchant   British Trust for Ornithology, The Nunnery, Thetford, Norfolk IP24 2PU

P. Millett   School of Applied Sciences, University of Wolverhampton, Wulfruna Street, Wolverhampton WV1 1SB

A.J. Moffat   Environmental Research Branch, The Forestry Authority, Alice Holt Lodge, Wrecclesham, Farnham, Surrey GU10 4LH

J.R. Packham   School of Applied Sciences, University of Wolverhampton, Wulfruna Street, Wolverhampton WV1 1SB

G.S. Patterson   The Forestry Authority, 231 Corstorphine Road, Edinburgh, EH12 7AT

G.F. Peterken   Beechwood House, St Briavels Common, Lydney, Gloucestershire, GL15 6SL

J.S. Rodwell   Unit of Vegetation Science, University of Lancaster, Bailrigg, Lancaster, LA1 4YG

M. Sangster   The Forestry Authority, 231 Corstorphine Road, Edinburgh, EH12 7AT

I.F. Spellerberg   Centre for Resource Management, PO Box 56, Lincoln University, Canterbury, New Zealand

J.W. Spencer   English Nature, Foxhold House, Crookham Common, Newbury, Berkshire, RG15 7EE

I.C. Trueman   School of Applied Sciences, University of Wolverhampton, Wulfruna Street, Wolverhampton WV1 1SB

Foreword: An Overview of Native Woodland Creation

G. F. PETERKEN

INTRODUCTION

New woodlands are being created all around us. From a low point of 4.5% of the land area in 1895, the area of woodland in Britain has grown to 11%, and will continue to grow. The great majority of these new woodlands are plantation forests of Sitka spruce (*Picea sitchensis*) and other introduced conifers in the uplands of the north and west and there have also been significant additions in the lowlands, for example, in Thetford Chase, the Suffolk Sandlings and parts of the South Downs. For decades these forests were largely ignored by ecologists, but, as they have matured and become more diverse, their ecological development has increasingly been studied. Several publications are now available which bring together much of this new information (Ford *et al.*, 1979; Jenkins, 1986).

The extent of native woodland has increased little, if at all, during this period. Large tracts of farmland reverted to woodland during the decades before 1940, but most of this woodland has since been reclaimed by agriculture. More recently, commons and marginal pastures have been neglected, with the result that many have succeeded to woodland, but over the same period 8% of all ancient woodlands have been destroyed and it is doubtful if there has been a net gain.

Precedents exist for 20th century afforestation which furnish an opportunity to understand the long-term ecological development of new woodlands. New native woodlands must have been a common feature of the post-Roman and late medieval landscapes. Small plantations of mainly native trees and shrubs were common in the 18th and 19th centuries as open fields gave way to designed landscapes containing scattered coverts and belts (see Bell, Chapter 3). The Board of Agriculture reports record that conifers were commonly planted in the early 19th century, pure or in mixture with broadleaves (Jones, 1961).

This introduction concentrates on the ecological case for enlarging the native woodland area of Britain. Although the new woodlands which are the subject of this book are by no means restricted to native woods, the introduction provides a context for many of the chapters that follow.

Definition: what is native woodland?

'Native woodland' is not amenable to a single, precise definition. It certainly includes all woodland composed of locally native tree and shrub species which have not obviously

The Ecology of Woodland Creation. Edited by Richard Ferris-Kaan.

© 1995 The editor, contributors and the Forestry Authority. Published in 1995 by John Wiley & Sons Ltd.

been planted, i.e. semi-natural woodland. Some ecologists would restrict it further to woods composed mainly of trees of local provenance, established by natural regeneration, but this would exclude such obviously 'natural' woods as the oakwoods on Loch Lomondside.

Precise definition has no virtue outside census methodology and the need for fair and unambiguous application of forestry incentives and controls. For general purposes there is merit in recognising different degrees of nativeness, which accept the importance of local genotypes and natural regeneration but also admit stands composed of species translocated from other parts of Britain and, in certain circumstances, other species which have become naturalised. Table F.1 summarises a possible range of definitions.

Table F.1. Native woodland defined on a sliding scale according to origins of the species and stands. Each category would include the previous categories when referring to the total native woodland resource. Thus, the extent of native woodland for strict nature conservation purposes is less than the extent for general environmental purposes

	Species and genotypes locally native	Species locally native	Species nationally native, but not locally native	Species introduced to Britain
Stand naturally regenerated	*****	***	**	*
Stand planted	****	***	*	—

***** The strict definition, which can be further refined by insisting that regeneration must have been natural throughout the history of the wood, that the site has never been disturbed, etc.Suitable for nature reserves.

**** A practical version of the strict definition, admitting planting as a means of restocking, but accepting some genetic change due to selection and altered survivorship of planted stock. Suitable for managing ancient semi-natural woods where there is no certainty that existing stock was planted using distant sources.

*** A practical and reasonably strict definition, which would be acceptable for most conservation interests, especially if applied to species which have commonly been planted, such as beech (*Fagus sylvatica*) and oak (*Quercus* spp.).

** A broad definition, which would be acceptable to most conservation interests if it is applied to new native woodland, but not ancient semi-natural woodland.

* The broadest meaningful definition, which would probably be acceptable to the general public for amenity management.

Extent of native woodland

Some 85–90% of Britain was originally covered in woodland. Remnants of these original woods survive in the modern landscape as ancient, semi-natural woods. Recent estimates based on the Nature Conservancy Council Inventory of Ancient Woodlands, show that these remnants have been reduced to 318 000 ha, or 1.4% of the land surface (Spencer and Kirby, 1992; Roberts *et al.*, 1992).

The extent of secondary native woodlands is not known, but an estimate can be made by starting with the most recent Forestry Commission 1979–82 census results and making assumptions about the origin and composition of each category. On this basis, it is estimated that Britain contains 231 000 ha of secondary, semi-natural woodland growing on old farmland, heaths, moors, bogs, railway embankments, quarries and other formerly unwooded ground (K.J. Kirby, personal communication). The census itself recorded 249 822 ha of high forest stands of nationally (i.e. to England, Scotland or Wales) native tree species, of which 118 252 ha was Scots pine (*Pinus sylvestris*) in Scotland and 131 570 ha was broadleaved. The two estimates certainly overlap, but combined they total 300 000–400 000 ha of secondary native woodland. The total extent of ancient and secondary native woodland is therefore unlikely to be much more than 3% of the land surface. This is an order of magnitude less than the proportion in many European countries, which not only have more woodland, but also produce more of their timber from locally native trees.

The native woodlands are generally small. England and Wales, for example, possess 27 688 separate ancient woods, of which 83% are 20 ha or less, and only 500 extend to more than 100 ha (Spencer and Kirby, 1992). Secondary native woodland is just as thoroughly broken up into a scatter of small woods, surrounded by farmland. Inevitably, most individual woods are isolated from each other, the populations of wildlife species in one wood having little contact with those in other woods. This isolation has been intensified in recent decades by hedgerow removal, the loss of farmland trees, ditching of streams, and ploughing of old meadows and pastures, all of which have made farmland even more hostile to woodland plants and animals.

The need for more native woodland

In a country with such small and isolated native woodlands, there are many strong ecological reasons why we need more and larger native woods.

- To increase the populations of woodland species. This is best achieved by expanding existing woods, for woodland species can spread into new woodland without the hazards of crossing hostile farmland (Peterken and Game, 1984).
- To increase population resilience, i.e. to enable gene exchange between populations of woodland species, and enable local extinctions to be reversed by colonisation. This is best achieved by habitat links between existing woods (Hanski and Thomas, 1994).
- To establish more edge habitats and habitat mosaics. These are especially rich wildlife habitats, which cater for species requiring more than one habitat and those with large territories (Kirby, 1992).
- To improve the quality of other habitats, such as streams, which can be buffered by woodland against the environmental impacts of farming (Pinay *et al.*, 1993).
- To restore degraded soils to higher levels of biological productivity (Miles, 1986).

For all these purposes, woodlands composed of native trees and shrubs are generally better than woodlands of introduced trees. Outside the Highlands, where *Pinus sylvestris* is indigenous, native deciduous broadleaves are markedly superior to introduced, evergreen conifers.

Native woods are, of course, also important as sources of hardwood timber and small wood, and they can be much improved in this respect. They form an essential component of the cultural landscape, to the degree that the distinctive character of a district can be reinforced if new woodlands contain a substantial native component and reinforce the existing pattern of woodland (Price, 1993).

MAKING MORE NATIVE WOODLANDS

It is often assumed that the only method of creating new native woodlands is to plant native trees on unwooded ground, but this is unnecessarily restrictive. Native woodlands can be created by other means and there would be many ecological benefits if a broader view were taken of the matter.

Methods

It is convenient to recognise four methods of creating new native woodland, which can be combined in various ways:

1. Planting native trees on unwooded ground. Oak (*Quercus petraea* and *Q. robur*), ash (*Fraxinus excelsior*), beech (*Fagus sylvatica*), hazel (*Corylus avellana*) and other natives were commonly planted in the past. Recently, many lowland farmers have planted native trees in small patches in field corners. In 1962, the Nature Conservancy Council embarked on an ambitious programme of native woodland creation on Rhum (Plate F.1), and now there is not only an extensive developing forest, but sound information has been accumulated on the build-up of the woodland fauna (Wormell, 1977).

Plate F.1. New native woodland planting on Rhum, The Western Isles, Scotland (G.S. Patterson, The Forestry Authority)

2. Natural regeneration on unwooded ground. This was a common process during the agricultural depression of 1870–1940 (Tansley, 1939). Scrub woodlands are still frequently seen springing up on steep slopes, old commons and waste ground. In the Highlands, much of the existing native woodland arose in this way during periods of reduced grazing.

3. Adaptation of conifer plantation forests, especially in the uplands. Native trees and shrubs are steadily colonising these forests naturally. More are being planted in fulfilment of the policy of permitting 5% of the area to be broadleaved. The area of such forests is about 1.5 million ha (Forestry Commission, 1993), of which perhaps 1.3 million ha is upland conifer plantations. If 5% of this area, i.e. 65 000 ha, were allowed to be native woodland, this would be a very significant addition to the area of native woodland, *sens. lat.*

4. Restoring ancient woods which have been changed to plantations of introduced species. The Forestry Commission has, for example, removed conifers planted under a mature oakwood at Dalavich, Argyll, and is restoring the substantial parts of Chalkney Wood, Essex, to native mixtures where conifer plantings did not kill the preceding coppice. In Britain some 217 154 ha of ancient woodland is occupied by plantations of all kinds (Spencer and Kirby, 1992; Roberts *et al.*, 1992).

Location

The ground available for new forests depends on decisions made by individual landowners and managers, taken in the light of incentives and controls administered by various agencies. The overall pattern is shaped by strategic land-use policies, notably the Indicative Forestry Strategies and initiatives for a new National Forest in the North Midlands and Community Forests close to several cities (Goodstadt, 1991; Marsh, 1993). At a local scale, preferences are influenced by the need to maintain and improve the characteristic and natural features of the landscape.

The ecologist can only state general principles, in the expectation that decision-makers can devise methods of incorporating them in their policies and practices. From the ecological standpoint, two sets of issues are important:

1. The ecological value of the land which might be replaced by new woodland.
2. The pattern of new woodland which will facilitate the ecological development of any new woodland.

New native woods are highly desirable on existing arable, ley grassland and those upland pastures which lack important moorland plants and animals. These are all extensive habitats supporting a limited range of wildlife, where new native woodland would bring a net gain for wildlife. On the other hand, new woodland of any kind may be damaging on lowland meadows, upland mires, herb-rich pastures and heathlands, which are already rich in native wildlife and natural features, and which would be diminished or damaged by a change to woodland. In most districts these habitats have been reduced so much that we need to protect all remaining examples.

Pattern of new woodland

Any new woodland must be inserted into the pattern of woodland we inherit. Wherever it is placed, new woodland is likely to be beneficial ecologically — apart from the exceptions listed above — but the benefits would be greater if the distribution of new woodland were based on the following locational principles:

● Existing woods should be enlarged, preferably by natural regeneration. This will help to minimise local extinctions of wildlife species in ancient woods.

- Large tracts of moorland should be diversified, partly by creating new woodlands in the vicinity of semi-woodland habitats, i.e. cliffs, screes, outcrops, incised streams and bracken brakes, where many woodland species hang on outside woodlands.
- Links between scattered woods should be restored by recreating farmland habitats and by scattering small plantings between woods. Habitats which should be restored are not just hedges and farmland trees, but also corridors of semi-natural habitats along riversides and along steeper slopes.
- In the uplands, the native woodland being created within plantation forests should be linked to habitats outside the forests. In particular, native woodland should be associated with riparian corridors, both inside the forests and outside.

Choice of species

When new woodland is created by natural regeneration we effectively let nature select the species from the pool created by the history of land management in the immediate neighbourhood, but when new woodland is created by planting we make the selection for ourselves. Five species-selection principles can be recognised, which, if observed, should optimise the ecological development of new woods and minimise the risks of establishment failure (Soutar and Peterken, 1989; Rodwell and Patterson, 1994):

1. Mixtures of species are preferable to monocultures. They create a diverse habitat, afford flexibility for later treatments, and provide some insurance against failure of any one species.
2. The species should be native to the district. Locally appropriate mixtures will maintain the *genius loci*, or local distinctiveness, both visually and ecologically (Lucas, 1991).
3. Local genetic stock should be used. With the exception of *F. sylvatica*, British provenances of native trees tend to survive and grow better than imported stock (Worrell, 1992).
4. The species should be appropriate to the site and soil (Chapter 8). This helps the new woodland to develop into a facsimile of long-established and ancient woods, minimises mortality of transplants and maximises growth rates.
5. Some rare or infrequent native trees should not be planted, because their distributions have meaning. For example, small-leaved lime (*Tilia cordata*) and wild service (*Sorbus torminalis*) are indicators of ancient woodland. Others have scientific interest which would be impaired by planting, e.g. the micro-species of whitebeam (*S. aria*). These should only be planted under controlled and recorded conditions.

Naturalised species

Sycamore (*Acer pseudoplatanus*) presents many problems and guarantees disputes within and between conservationists and foresters. It is easily established and readily colonises existing woodlands, but it is not a native species. It is too widely and thoroughly established to be exterminated, even if the conservation case for doing so is strong, but there is a good case for controlling its tendency to dominate in some woodland types.

As a general rule, *A. pseudoplatanus* and other naturalised species should not be planted next to ancient woods which do not already contain it. Elsewhere it can be accepted in natural regeneration. It can also be planted where it is traditional, for example in the coastal zone, around upland farms and in urban forests (Boyd, 1992).

Each species has individual characteristics which should be considered on their merits. For example, *A. pseudoplatanus* brings some nature conservation benefits (Taylor, 1985) and can be accepted with qualifications. Rhododendron (*Rhododendron ponticum*), on the other hand, persistently casts very heavy shade: it is an unmitigated menace in ecological terms, and should be excluded.

Planting the woodland ecosystem

Planting trees does not by itself create a woodland. Woodland plants and animals must colonise from populations in existing woods and semi-woodland habitats. Soils must develop woodland characteristics. These processes continue through the first rotation and beyond. Colonisation remains incomplete even after several hundred years, and some species of ancient woodlands may never colonise new woodlands on similar sites (Peterken and Game, 1984; Peterken, 1993).

Where new woodlands are initiated close to an existing native wood it is preferable to allow native plants and animals to colonise naturally. Such new woods are not ecologically isolated, so the build-up of a woodland fauna and flora is reasonably assured. Where new woodland is established in isolation from existing woods, the barriers to colonisation may be partially overcome by inoculating the new wood with species appropriate to the site (Packham *et al.*, Chapter 9). In practice, the woodland flora can be assisted in this way, but the fauna will have to look after itself.

HOW MUCH NEW NATIVE WOODLAND DO WE WANT?

Britain

In a land so depleted of native woodland, there is little risk of producing too much, but specific targets are desirable. If the national area were doubled, Britain would still have only 6% cover of native woodland. In some especially depleted regions, such as the Highlands, we need perhaps 3–5 times the present area.

More important than an arbitrary national target, there are good ecological reasons for clustering the new woodland, rather than dispersing it evenly. In sparsely wooded districts new native woodland usually remains poor in woodland species, but in well-wooded districts, such as parts of the Weald, Chilterns, Welsh Borderland, Deeside and Speyside, new woods are often so well endowed with woodland species that they are almost indistinguishable floristically from ancient woods, i.e. in such districts it seems that ecological isolation is limited enough to allow woodland species to disperse and establish themselves in new woods. Such well-wooded districts are generally characterised by 25–30% woodland cover supplemented by numerous semi-woodland habitats, such as hedges. These conditions may prove to be a threshold at which a landscape is ecologically resilient for woodland species.

On this basis, the targets should focus more on generating new well-wooded districts with at least 30% woodland cover, containing a substantial fraction of native woodland. For maximum benefits these districts should be developed around existing woodland clusters. Thus, the best strategy would be to reinforce the existing pattern at scales from the sub-parish to county and regional.

Some experiences from other countries

In the eastern United States, a very substantial area of farmland has reverted to woodland in the last century. Most of this new woodland consists of mixtures of native trees and shrubs, which have developed naturally. In some regions where woodland cover was once reduced to 20%, it has now recovered to 90%, leaving only small, isolated pockets of farmland. Timber in the new woodland has been harvested and the woodland has been allowed to regenerate again by natural means, but very often this second generation woodland has differed substantially in composition from the post-farming pioneer woodland. While in some regions the new woodlands have more-or-less restored the pattern of forest types recognised by the first Europeans (Whitney, 1991), in others new forest types have been generated (Whitney, 1987). New woodland is unlikely to develop on this scale in Britain, and if it does the survival of non-woodland habitats will become a conservation priority, but it is useful to bear in mind that any new native British woodland will probably include new semi-stable combinations with *A. pseudoplatanus, Fraxinus excelsior* and *Fagus sylvatica*.

The remnants of original or primary woodland remain floristically distinct in the eastern United States, even though they are now embedded in a matrix of woodland (Whitney and Foster, 1988). In fact, ancient woodland indicator herbs have now been identified in the forests of many European countries in addition to Britain, such as Belgium (Tack *et al.*, 1993), the Czech Republic (Kubikova, 1987), Germany (Zacharias and Brandes, 1990), Poland (Dzwonko and Loster, 1988) and Sweden (Brunet, 1993). The limited capacity of saproxylic species to spread from refuges is also well recognised and widespread (Speight, 1989; Rose, 1988). The importance of habitat continuity and the need to connect new woodlands with old woods and semi-woodland habitats is increasingly recognised, but so too are the natural limits on the rate at which biodiversity can develop in new woodlands.

SUMMARY

The time is now right for a fresh look at new woodlands and their ecological development. UK forestry policies are no longer restricted to wall-to-wall spruce afforestation and there is widespread support for creating new woodlands with a strong native element which are designed to fit in with the existing local landscape. At the same time, two decades of ecological research have yielded a deeper understanding of how and at what rate these new woodlands develop as ecosystems, and what their ultimate limitations and values may be.

REFERENCES

Boyd, J.M. (1992). Sycamore — a review of its status in conservation in Great Britain. *Biologist*, **39**, 29–31.
Brunet, J. (1993). Environmental and historical factors limiting the distribution of rare forest grasses in south Sweden. *Forest Ecology and Management*, **61**, 263–275.
Dzwonko, Z. and Loster, D. (1988). Species richness of small woodlands on the western Carpathian foothills. *Vegetatio*, **76**, 15–27.

Ford, E.D., Malcolm, D.C. and Atterson, J. (eds) (1979). *The Ecology of Even-aged Forest Plantations*. Institute of Terrestrial Ecology, Cambridge.

Forestry Commission (1993). Forestry facts and figures 1992–93. Forestry Commission, Edinburgh.

Goodstadt, V. (1991). Indicative forestry strategies — the Scottish experience. In: Millan, C. and Maloney, M. (eds). *The Right Trees in the Right Place*. Seminar Proceedings 4. Royal Dublin Society, Dublin, 108–115.

Hanski, I. and Thomas, C.D. (1994). Metapopulation dynamics and conservation: a spatially explicit model applied to butterflies. *Biological Conservation*, **68**, 167–180.

Jenkins, D. (ed.) (1986). *Trees and Wildlife in the Scottish Uplands*. Institute of Terrestrial Ecology, Banchory.

Jones, E.W. (1961). British forestry in 1790–1813. *Quarterly Journal of Forestry*, **55**, 36–40 and 131–138.

Kirby, P. (1992). *Habitat Management for Invertebrates: A Practical Handbook*. Royal Society for the Protection of Birds, Sandy.

Kubikova, J. (1987). Cultivated forest stands in central Bohemia, their floristic composition and history. Martin Luther University Halle Wittenberg. *Wissenschaftlichebeitrage*, **46**, 155–165.

Lucas, O.W.R. (1991). *The Design of Forest Landscapes*. Oxford University Press, Oxford.

Marsh, S. (1993). *Nature Conservation in Community Forests*. Ecology Handbook 23. London Ecology Unit, London.

Miles, J. (1986). What are the effects of trees on soils? In: Jenkins, D. (ed.). *Trees and Wildlife in the Scottish Uplands*. Institute of Terrestrial Ecology, Banchory, 55–62.

Peterken, G.F. (1993). Long-term floristic development of woodland on former agricultural land in Lincolnshire, England. In: Watkins, C. (ed.). *Ecological Effects of Afforestation*. CAB International, Wallingford, 31–43.

Peterken, G.F. and Game, M. (1984). Historical factors affecting the number and distribution of vascular plant species in the woodlands of central Lincolnshire. *Journal of Ecology*, **72**, 155–182.

Pinay, G., Roques, L. and Fabre, A. (1993). Spatial and temporal patterns of denitrification in riparian forest. *Journal of Applied Ecology*, **30**, 581–591.

Price, G. (1993). *Landscape Assessment for Indicative Forestry Strategies*. Forestry Authority, Cambridge.

Roberts, A.J., Russell, C., Walker, G.J. and Kirby, K.J. (1992). Regional variation in the origin, extent and composition of Scottish woodland. *Botanical Journal of Scotland*, **46**, 167–189

Rodwell, J. and Patterson, G. (1994). *Creating New Native Woodlands*. Forestry Commission Bulletin 112. HMSO, London.

Rose, F. (1988). Phytogeographical and ecological aspects of Lobarion communities in Europe. *Botanical Journal of the Linnean Society*, **96**, 69–79.

Soutar, R.G. and Peterken, G.F. (1989). Regional lists of native trees and shrubs for use in afforestation schemes. *Arboricultural Journal*, **13**, 33–43.

Speight, M.C.D. (1989). *Saproxylic Invertebrates and Their Conservation*. Council of Europe, Strasbourg.

Spencer, J.W. and Kirby, K.J. (1992). An inventory of ancient woodland for England and Wales. *Biological Conservation*, **62**, 77–93.

Tack, G., Bremt, P. van den and Hermy, M. (1993). *Bossen van vlaanderen*. Davidsfonds, Leuven.

Tansley, A.G. (1939). *The British Islands and Their Vegetation*. Cambridge University Press, Cambridge.

Taylor, N.W. (1985). *The Sycamore in Britain — Its Natural History and Value to Wildlife*. Discussion Papers in Nature Conservation 42. University College, London.

Whitney, G.G. (1987). An ecological history of the Great Lakes forest of Michigan. *Journal of Ecology*, **75**, 667–684.

Whitney, G.G. (1991). Relation of plant species to substrate, landscape position, and aspect in north central Massachusetts. *Canadian Journal of Forest Research*, **21**, 1245–1252.

Whitney, G.G. and Foster, D.R. (1988). Overstorey composition and age as determinants of the understorey flora of woods of central New England. *Journal of Ecology*, **76**, 867–876.

Wormell, P. (1977). Woodland insect population changes on the Isle of Rhum in relation to forest history and woodland restoration. *Scottish Forestry*, **31**, 13–36.

Worrell, R. (1992). A comparison between European continental and British provenances of some British native trees: growth, survival and stem form. *Forestry*, **65**, 253–280.

Zacharias, D. and Brandes, D. (1990). Species–area relationships and frequency — Floristical data analysis of 44 isolated woods in northwestern Germany. *Vegetatio*, **88**, 21–29.

Acknowledgements

I am indebted to many people who, by attending the British Ecological Society's Forest Ecology Group Symposium *Ecological Aspects of the Creation of New Broadleaved Woodland* in Leicester in April 1993, encouraged me to synthesise the themes of the meeting into this book. I wish to thank the many original contributors to the symposium who have submitted modified papers, and others who generously agreed to write additional material. For their invaluable help in the editorial process, I thank Jenny Claridge (advice and copy editing), John Williams (artwork), Jill Shipp and the staff of the Typing Pool, Alice Holt Lodge (typing the original manuscript). Lastly, I wish to thank Alison for her patient support throughout the lengthy editorial process.

The support of the British Ecological Society in funding meetings of the Forest Ecology Group is acknowledged, without which many useful exchanges of information and views could not have taken place.

Richard Ferris-Kaan
August, 1994

1 To What Extent Can We Recreate Woodland?

J. W. SPENCER

INTRODUCTION

> There are fashions in land-use, and they tend to generate fashions in landscape, but there is an important difference from other types of fashion. If one orders a fashionable dress, one expects to wear it before it becomes out of date. But landscapes depend on the growth of trees and other long-lived plants, and can never catch up with a fashion before it changes. (Rackham, 1991)

Any habitat creation project, but especially any woodland habitat creation project, is destined to mature in ways unforeseen by its planners. Such projects ultimately give rise to results that are unlikely to be seen, let alone planned for, by their originators. Their singular feature is that their maturation time is very much longer than the working span of the individuals involved in them. In woodlands it is excessively longer. Some essential woodland habitats, such as maturing deadwood, may take centuries to arrive, let alone develop into species-rich ecosystems. Given that the very value of most habitat creation projects lies in their long-term, indefinite contribution to the conservation of organisms associated with them, it is essential that conservation planning accommodates for these human failings from the outset if it is to achieve its goals.

In this chapter an attempt has been made to identify ways in which we might make woodland habitat creation projects 'fashion resistant'. How might we ensure that our original primary objective, that of allowing a species-rich, more or less natural woodland to develop over many decades, has at least some chance of success in the face of continuing change in society at large? It is an attempt to consider the necessary human aspects of woodland creation, alongside the more obvious ecological aspects.

CREATION VERSUS RECREATION

There are several key issues that need to be addressed when setting out any long-term woodland project. Our present understanding of these issues, and the frailty of human involvement in woodland projects throughout history, relies heavily on the work of Rackham (1980, 1986b, 1991), Peterken (1993), and Backmeroff (Peterken and Backmeroff, 1988). The main themes that emerge can be expressed as three key lessons.

The Ecology of Woodland Creation. Edited by Richard Ferris-Kaan.

Lesson 1

Attempts at creating functioning woodland habitat are fated to success. We know this simply because so many forests and woodlands, on a wide range of soils and in a wide range of circumstances, have reassembled themselves into woodland ecosystems. Indeed the majority of the highly valued ancient semi-natural woodlands in England may be secondary in origin (Taylor, 1983). Many of these secondary semi-natural woodlands are valued for a wide range of reasons, including nature conservation.

Lesson 2

Attempts at recreating ancient woodland or any other ancient habitat in Britain are doomed to failure. Ancient woods are, by definition, the products of long periods of existence set in a particular past (Peterken, 1993). The term implies a long continuity of woodland cover for present-day woods and forests that have been profoundly influenced by medieval woodland management practices of one sort or another. A woodland coming into being now may accrue age and, in a couple of centuries, become ancient within the general use of the word, but it cannot become an ancient wood of medieval origin. The past origins, management and surroundings cannot be replicated. We do not know or understand enough to replicate the ecological and human conditions that prevailed when such places were created. The conditions under which they evolved — the past climates, surrounding landscapes, management practices, etc. — are certainly not replicable, let alone the long periods of time over which they have exerted their influence. We do know that the long periods of time spent under particular management regimes were very important elements in their creation, but we cannot embark on a century of coppice work or meadow mowing with any real conviction that we shall be able to continue for more than a few years. No human society can reasonably set out to pursue a project or experiment that is not expected to come to fruition for centuries to come, and no human society should set out on a project based on the illogicality of recreating that which it knows to be unrecreatable. At best such attempts create a pastiche of their models, devoid of the meaning of the originals they copy.

Lesson 3

Attempts at creating woodlands are unpredictable in outcome. We can create woods, and greatly influence their character, composition and value, but we must not expect to be too accurate in predicting the details of the final result. We do however have enough insight into woodland processes and succession to design the starting conditions that will lead to the development of native woodland appropriate to the site and its conditions, and likely to fulfil the broader objectives that led to its establishment.

HABITAT CREATION PROJECTS

The historical lessons of woodland management

History is a powerful force, shaping both the present and the future. Nowhere is this more evident than in the woodlands of the present-day landscape. In recent years many ecologists have discovered that historical sources are an essential tool in the study of woodlands (e.g. Rackham, 1980), as important to their understanding as studies of ecological processes, soil chemistry or the interaction of species. Historical studies have not only given us a greater insight into the past impact we have had on woodlands through their exploitation and management, they have also informed us about the nature of human endeavour within woods and forests, and of our varying commitment to projects whose fruition lies well into the future. The life of this commitment is seriously influenced by changes in fashions of both land management and scientific interest.

Rackham (1991) has written at length on the importance of conserving the meaning of the historical landscape, illustrating the extent to which fashion and short-lived enthusiasm, coupled with neglect and default, have shaped the woodland landscape. This essay is essential reading for anyone embarking on a long-term exercise aimed at establishing woodland, whatever their motives for its creation. It exposes many of the great human weaknesses inherent in such grand visions. 'The countryside is full of monuments to successive fashions in forestry of the last 250 years. People planted trees and either forgot to cut them down when the time came or left them standing because the use for which they had been planted had disappeared by the time the trees had grown' (Rackham, 1991).

In Britain the last two centuries have seen some major changes in the fashions for tree planting and woodland management. For example, it was once the height of tree planting fashion to plant Weymouth pine (*Pinus strobus*). How many are planted today? Extensive areas of oak (*Quercus* spp.) were planted to cater for the perceived needs of the Royal Navy (Rackham, 1980). Most have only become suitable for shipbuilding in this century. There was once a passion for under-planting native hardwoods with conifer nurses; many such areas are now being prematurely removed from ancient woods across the country. More recently there has been an enthusiasm for placing sculptures in woods and forests across the land. The dominant fashion today is a passion for ancient semi-natural woodland and a return to traditional management. This too is now beginning to wane in favour of an interest in more natural stands of mature old growth native forest. How long can we expect these fashions to last? What most have in common is that they arrive, wax strong, wane and almost disappear, well within the lifespan of even the shortest-lived trees and shrubs.

Clearly, there are important lessons to be learned from the study of woodland history. We shall not escape the pitfalls illuminated by these lessons if we fail to acknowledge them in our own projects. To achieve our objectives in establishing native woodland we therefore need to accept that we are subject to these influences and harness them to our own ends.

Long-term commitment

Projects involving long-term ecological change are peculiarly prone to changes in the circumstances that surround them, for example:

- *Changes in personnel*: new staff have new interests, new approaches and may be wholly unfamiliar with the effort put into establishing a long-term project a year or two before their arrival.
- *New technology and understanding*: this can have a devastating effect on long-term

projects; there is always a loss of interest in continuing with a project set up using outmoded techniques, recorded on outmoded software (such as paper cards!), and set up to answer problems long since considered irrelevant.

● *Project funding*: where this is reduced opportunities for pursuing projects as planned can be severely curtailed. Conversely, increased funding can broaden opportunities to the point where the original objectives may be lost. All long-term projects are prone to 'funding fatigue', diminishing the support of funding bodies as the initial interest and excitement wanes.

INFLUENCES ON THE DEVELOPMENT OF NEW NATIVE WOODLAND

Landscape is created not only by human design but by human default. This is still not fully taken into account. (Rackham, 1991)

American foresters and ecologists are well aware of the ecological consequences of human default. In New England alone, after 400 years of endeavour to create hard won farmland, an area the size of the British Isles has turned from farmland into woodland over the past 200 years. The forests of North America are now nearly 80% as extensive as they were in 1600 (Williams, 1989; Figure 1.1). Much of this forest is rich in history, ecologically interesting and biologically diverse, being the natural response of the forest ecosystem to disturbance and neglect at varying times in the past, in varying places and on varying scales (Plate 1.1). A large proportion of this forest area now contributes to a major industry, producing high quality timber. Such places reveal much about woodland habitat creation.

Woodland can, and readily does, successfully create itself, with only exceptionally high, dry, rocky or regularly disturbed areas failing to become largely dominated by trees and shrubs of one sort or another. Such areas, along with those that are regularly browsed or grazed by native forest animals, would have had a long continuity as open conditions and would have supported the 'non-woodland' species element within the natural forest.

Natural colonisation

This same spontaneous recreation of large secondary forests has occurred in many other parts of the world. In Yucatan large areas of secondary forest now form part of an extensive forest biosphere reserve of international importance (Batisse, 1986; SEDUE, 1987). Most of this area was under Mayan agriculture until a few centuries ago. The archaeological remains add to the meaning of the biosphere reserve and its importance, rather than detracting from it. In North Borneo, along the lower stretches of the Kinabatangan River, there lies one of the last great remaining tropical riverine forests in South East Asia. This forest is rich in plants, birds and insects, along with rare mammals such as the Sumatran rhinoceros, Asian elephant, proboscis monkey and orangutan. However, large stretches of this forest, presently being hard fought for by Malaysian conservationists, are situated on the sites of old Dutch tobacco plantations abandoned sometime in the 1850s (Payne, 1990 and personal communication).

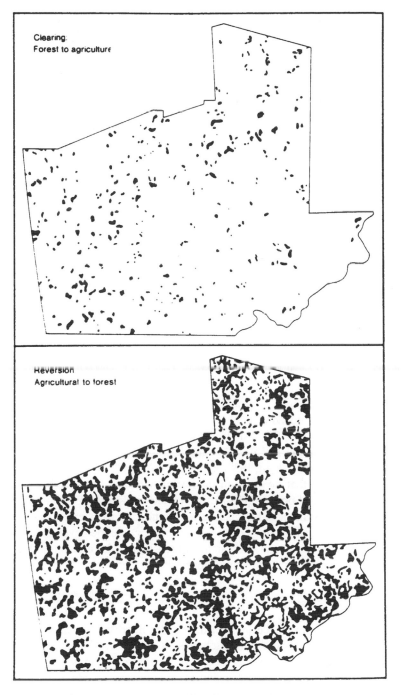

Figure 1.1. Conversion of forest to agricultural land and abandonment of agricultural land to secondary forest, Carroll County, Georgia, USA, 1937–1974 (J.F. Hart, *Land Use Change in a Piedmont County*. Reproduced from M. Williams (1989), *Americans and their Forests: A Historical Geography*, by permission of Blackwell Publishers

Plate 1.1. Regeneration of woodland on abandoned agricultural land, Petersham, Massachusetts, USA. The old wall marks the original field boundary, and the sugar maple *(Acer saccharum)* would have been planted as a 'farmland' tree, grown for its utility (J.W. Spencer, English Nature)

Within England there are many important woods noted for their wildlife interest that have arisen on past agricultural landscapes; Wychwood Forest, a National Nature Reserve in Oxfordshire, lies on the site of extensive Roman farmland settlements (Emery, 1974); Bowdown Woods in Berkshire, an important nature reserve managed by the Berkshire, Buckinghamshire and Oxfordshire Naturalists Trust, is partially on the site of an abandoned Roman villa (personal observations); Overhall Grove, an ancient wood in west Cambridgeshire, has developed on the site of the deserted medieval village of Knaphill abandoned sometime in the 15th century (Peterken, 1993). Even the 19th century plotland woods of Basildon and Rochford are interesting enough to warrant their management as a Country Park (Rackham, 1986a).

Since it is so evident that diverse and interesting woodland can readily arise spontaneously without much call on the professional services of foresters or ecologists, what is their role in the establishment of new native woodland? To what extent can their understanding of forest composition and dynamics be used to guide the development of biologically rich, functioning woodland ecosystems which will make a lasting contribution to the conservation of native plant and animal species?

Time

Time is of paramount importance in determining the interest and value of any tract of woodland for nature conservation. Invertebrates of glades and open spaces, and other habitats of early succession, can be catered for very early on in the development of a new native woodland (see Key, Chapter 10), but invertebrates of deadwood habitats, heartwood fungi, the organisms of well-developed forest soils (see Harris and Hill, Chapter 7), and even many common bird species such as woodpeckers and nuthatches (*Sitta europaea*), cannot be accommodated until the habitat has matured over many years. Little can be done to accelerate the development of these mature woodland habitats on any reasonably extensive scale. Consequently, time and patience are absolute requirements if new native woodlands are to provide for many organisms associated with woodland. Securing this commitment to time is, without doubt, the most difficult of all issues to face when setting out on a long-term endeavour such as the creation of mature native woodland.

The problems experienced by Peterken and Backmeroff (1988) in setting up a series of long-term woodland studies clearly illustrate some of the difficulties to be encountered. Setting up their transects in various woods across the country, both before and after the storms of 1987, they commented that the sites for such long-term studies should be located in woods with an assured future as undisturbed nature reserves. National Nature Reserves were quoted as examples of such sites. The records made were to be lodged with institutions with a 'long-term view' of such studies. The Nature Conservancy Council (NCC) was cited as just such an institution, providing the necessary continuity of ownership and outlook needed to support such projects. Less than five years later the NCC had been split into four new organisations, major changes in staff had taken place, and the perceived role of the organisation, particularly with respect to woodland conservation, radically reinterpreted (Peterken, 1993). At least one NCC office was unable to locate the records of the study, lodged in its library only months before, and on enquiry the staff were unsure as to the nature of the work. While the sites themselves, and the transects within them, have to date remained sheltered from these changes, these examples vividly illustrate the vulnerability of long-term studies to changes in the nature of human institutions.

One vitally important recommendation emerges. When a serious decision has been made about creating old growth woodland — with or without various forms of management — steps should be taken at the outset to legally tie up the ownership and management of the land. These legal ties should be designed to ensure that the aims and objectives are both closely defined from the beginning and implemented well into the future. Management, whether via grazing or the exploitation of wood and timber, may need to be vested in self-interested parties that not only stand to gain from the activity, and hence ensure its continuity, but will jealously guard the right to do so. The New Forest in Hampshire stands as a striking testimony as to how successful such rights can be in safeguarding continuity (Tubbs, 1986).

Legal covenants, the establishment of trusts and the setting up of modern-day commons could all be used as implements to influence future management. Restrictions on the use of non-native species, undesirable management practices, the establishment of grazing rights (in wood pasture projects) or the extent of planting once the establishment of the wood was completed, could all be considered in the articles drawn up for such covenants and trust. While not wholly buffering such projects from future change, such legal implements would go a long way to fashion-proofing woodland creation projects and ensuring their long-term future continuity.

Extent

Compared to long periods of time, large expanses of ground are comparatively easy to assemble! Nevertheless they remain a major obstacle to consider when planning a woodland creation project. Several observations might be made:

- It is clear from many studies of woodland that the more extensive the tract the more likely it is to contain a mosaic of woodland types and support a diverse flora and fauna

(Peterken and Game, 1984). It is also more likely to support sufficiently large populations of rarer organisms, should they occur, for them to have some sort of assured future at the site. This is more of a general trend than a rule (see Spellerberg, Chapter 4). The Wyre Forest, for example, has some very large tracts of rather species-poor, rather uniform semi-natural oak woodland; by comparison Wykery Copse in Berkshire harbours more variation in woodland plant communities in 3 hectares than parts of the Wyre does in 300. Nevertheless, the larger the better has to be a simple and accurate axiom to follow in woodland creation projects.

● Areas where sufficiently large tracts of land could be obtained for really significant woodland creation projects are unlikely to be in regions that already contain any significant amount of existing ancient or long-established woodland. The experience of the Forestry Commission since the 1930s largely bears this out, as do the difficulties experienced by the Countryside Commission in attempting to establish Community Forests in the lowlands of England. There are some important exceptions in Scotland, where major opportunities for establishing extensive native pinewoods, within the range of existing relict pinewoods, have been recognised and are being enthusiastically and imaginatively pursued (Forestry Authority, 1994).

● Size also plays its part in buffering against changes in woodland management. Many large woods have retained species in the face of changes in composition or management compared to smaller woods. The survival of many woodland moths and butterflies in woodland converted to conifer plantations in the 1950s and 1960s is an example of this. Many of these woods also retain a little of the native woodland that predated these major changes. Decline in management activity would also appear to be buffered by larger sites. In many large semi-natural coppices there is still a little coppice work; the larger wood pastures and commons tend to be those that have retained some active commoners who continue to graze stock in these areas. The size of new woods may be an important determinant in ensuring future continuity of purpose.

ECOLOGY AND MANAGEMENT

Our understanding of woodland ecology and dynamics is probably sufficient for us to enhance the slow natural development of woodland, creating the conditions necessary for the development of biologically rich and interesting woods. We know, for example, that nitrogen-enriched soils produce rather dull and monotonous woods dominated by aggressive perennials; that thin, nutrient-poor soils give rise to much greater variation in species composition and structure in developing secondary woodlands; that soil heterogeneity generates inherent variation; that natural regeneration produces a more meaningful and appropriately placed distribution of trees and shrubs than planting (see Rodwell and Patterson, Chapter 5); that the provision of opportunities for recruitment and colonisation, through regular disturbance and disruption over longer periods of time, generates complexity in both structure and composition; that the presence of grazing animals can promote marked regeneration of trees in developing woodland (and that such woods have an interesting and complex structure); and that the structure of woods grazed by cattle is very attractive to a wide range of invertebrates. We can use this sort of information to guide soil or site preparation prior to tree planting, and the subsequent easing off of management activities (see Moffat and Buckley, Chapter 6). The National Vegetation Classification (Rodwell, 1991) provides an invaluable reference source for this sort of instructive material.

All this information can be used to guide and inform the decisions that need to be made at the outset of a woodland restoration project and during its subsequent development, either in the first few years of the project or, if we are really ambitious, over the coming centuries! The following could all be considered:

- Lopping, felling and coppicing could all be utilised in the first decades of a project's life, to generate structural and compositional complexity, and to provide continued opportunities for colonisation and recruitment, even within woods ultimately destined to become undisturbed old growth woods in the future.

- An approach not yet adopted by any woodland transplant project, but which looks an obvious avenue to explore given observation of woodland succession, would be to establish stands of fast growing, early pioneer trees like birch (*Betula* spp.) to create the woodland environment, into which other trees and shrubs and plants are subsequently introduced. Its fast growth, leaf fall and early senescence into deadwood (compared to other species) would do much to rapidly establish the soil and canopy conditions upon which many woodland organisms depend.

- The felling and deliberate damage of existing trees, winching some over to expose root plates, creating canopy gaps and deadwood habitat, are options for increasing naturalness in existing older stands. These options are evidently not open to managers of newly created woods for four or five decades, by which time the original objectives are almost certain to have been forgotten, but may be adopted where woodland habitat is being created from established stands. This approach is being seriously considered at Black Park Local Nature Reserve (Buckinghamshire County Council, 1991). Here, within a stand of mature oak (*Quercus* spp.) planted some 150 years ago on old heathland, the aim is to create a facsimile old growth stand, complete with fallen trees and upturned rootplates, for study by schoolchildren and other students, and comparison with other types of woodland within the surrounding Country Park.

Planting decisions

The great majority of woodland creation schemes accept without question the need to plant appropriate trees and shrubs. Few put as much consideration into the planting or establishment of other elements of the woodland flora and fauna. While we do not yet know enough about the role and function of many important woodland species, notably soil organisms (see Harris and Hill, Chapter 7), invertebrates and fungi, we do know enough to know that they are extremely important, and that we should consider inoculating new woodland ecosystems with a wide range of these organisms. Inoculation from existing woodland might be regarded as a perfectly valid course of action and may have an important role in the development of the desired native woodland ecosystem.

There is an argument that newly established woods would be most meaningful (i.e. tell us most about the conditions under which they have developed) if nothing, including the trees, were planted. Naturally established secondary woods are both more biologically diverse and more informative and meaningful than is generally appreciated. Once planting takes place, and the opportunities for natural development and change are curtailed, their potential value and interest declines rapidly. Restraint in planting is a key issue when considering the value and meaning of woodland projects being designed at present. While the no planting/no sowing viewpoint is wholly valid if the objectives are the study and appreciation of colonisation and succession, other equally valid approaches could be adopted in other circumstances. An enhanced conservation value might be reached much more quickly if trees, shrubs and other organisms were introduced in a scheme, even if the policy were to become one of non-intervention.

The issue of introductions becomes very complex very quickly, largely because the objectives at any one site are rarely clearly stated and also because attention rapidly focuses on the rare and unusual. Many people hold very strong views about the planting of rare species around the countryside, or the establishment of colonies of rare insects or birds. Much of the opposition is emotive and not clearly articulated, but there are also many strong logical or scientific arguments against widespread plantings and introductions (e.g. Joint Committee for the Conservation of British Insects, 1986; Maunder, 1992).

Consequently, it is important to consider the planting or introduction of organisms as two quite distinct classes of operation. The first and largest category would be the introduction of typical or cornerstone elements of woodland vegetation (e.g. hazel (*Corylus avellana*), hawthorn (*Crataegus monogyna*), bluebell (*Hyacinthoides non-scripta*) or dog's mercury (*Mercurialis perennis*)), soil organisms and the fungi of the appropriate woodland type. Promoting the establishment of a functioning woodland ecosystem would allow subsequent natural processes of colonisation, establishment and change to proceed apace. The second would be the introduction of locally rare or scarce species for which there is a need for conservation action and for which the woodland in question is deemed a suitable recipient site. The primary aim is to provide an extensive area where natural processes can dictate the pace and nature of change, with thoughtful management regimes designed to maximise the complexity and interest of the developments taking place.

Structure and composition

Much has been written about the structure and composition of native woodland and many practitioners have considerable insight into the combination of trees, shrubs and herbs that 'work' together in particular circumstances (Rodwell, 1991; Peterken, 1987; Soutar and Peterken, 1989). Two sources will provide most of what will be required for the foreseeable future.

● The National Vegetation Classification (Rodwell, 1991) provides a very powerful tool for guiding woodland design and predicting the consequences of woodland succession under different conditions (see Rodwell and Patterson, Chapter 5). It provides essential information on the keystone species of herbs and shrubs, as well as the trees, for the full range of British woodland types. An illustration of its use in woodland creation projects has been given in Rodwell and Patterson (1994).

● Studies on the dynamics of temperate forest will give us a similarly powerful framework for considering the importance of structural change, succession and disturbance in established woods and, by inference, an insight into how we might influence our woodland in the first few decades of their existence to maximise their 'naturalness' at the earliest opportunity.

THE CONSERVATION OF MEANING

Much of the value of our rare plants and animals lies not in themselves, but in what they can tell us. Their ecology, their past and present-day distributions, and their behaviour under different circumstances, all tell us a great deal about past climatic conditions, changes in land use, human activities and the response of the natural world to these changes. Lowland Britain is of particular importance in the study of the history of man's impact on the natural world (see Bell, Chapter 3). The consequences of our activities on the natural world have been considerable, over very long periods of time, but our historical records, the sophistication of our archaeological science, and the detailed knowledge we have of the ecology, distribution and history of our fauna and flora are exceptional on a worldwide scale. The conservation of the ecological and historical meaning of the countryside in England is not just a question of enjoyment and appreciation; it is a question of international scientific importance.

The conservation of meaning is a difficult concept to describe. Most of the meaning of a site lies hidden waiting to be discovered and interpreted. Once discovered, its importance may remain unappreciated by land managers and conservation biologists alike. Major difficulties lie in the fact that most sites have many layers of existing meaning. Attempting to ensure that newly established woods conserve their future meaning is probably an impossibility! Yet it is certain that given time, all of the woods established over the past 50 years, and all those to be planted in the coming 50, will eventually accrue meaning. It may be that they only reflect the planting fashions of the times; it may be that they become rich in important lessons for woodland ecologists. We cannot decide in advance what this meaning might be: it is too elusive an objective to worry about defining, but we can provide every opportunity for natural processes to proceed that will imbue the site with interest, worth and a sense of place.

CONCLUSION

The reasons for establishing a woodland need to be clear from the outset. A clear understanding of the objectives will readily determine what is and is not advisable in any particular project. Above all, arbitrary decisions regarding tree species choice should be avoided. Native woodland might be established to promote nature conservation, to enhance the future conservation prospects for a particular species, for timber production, sport, recreation or study. They might be planted for the enjoyment of nature in general or spring flowers in particular; for landscape, recreation or scientific study. All are valid reasons for establishing woods, and all options may be justifiably pursued somewhere in the country, but all are long-term projects subject to the vagaries of passing time. The extent to which it is necessary to worry about continuity and change in fashion varies with the objectives. Though each objective gives different meaning to the woodland it creates, not all are as dependent on the passage of time and commitment to a particular management regime to fulfil their objectives as those concerned with nature conservation.

The 'near natural' and 'native' in woodland are currently very fashionable. A few years ago it was 'traditional' management. In a few years it may be an enthusiasm for the planting of endangered tree species or plantations for biofuel. The only certainty is that these fashions will change faster than the trees can grow. Nevertheless the present fashion for

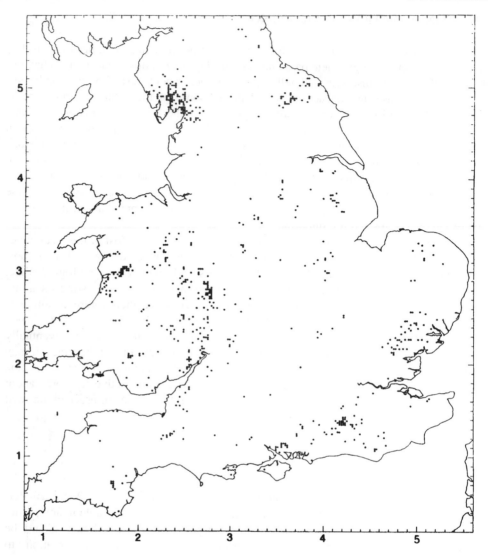

Figure 1.2. Native distribution of small-leaved lime *(Tilia cordata)* in the British Isles. The marginal numbers are the 100-km coordinates of the National Grid. Reproduced by permission of Blackwell Science Ltd from C.D. Pigott (1991), *Journal of Ecology*, **79**, 1147–1207

The planting of native lime trees (*Tilia cordata* and *Tilia platyphyllos*), or wild service tree (*Sorbus torminalis*) well illustrates this need to conserve meaning. All three species are rare trees and much of their meaning and interest lies in their past and present distributions (Pigott, 1991; Roper, 1993). Their idiosyncratic and interesting distributions are themselves in need of conservation. Rackham's comments on small leaved lime are very pertinent. 'The small leaved lime (*T. cordata*) is a relatively rare tree which hitherto has not usually been planted. It has now become

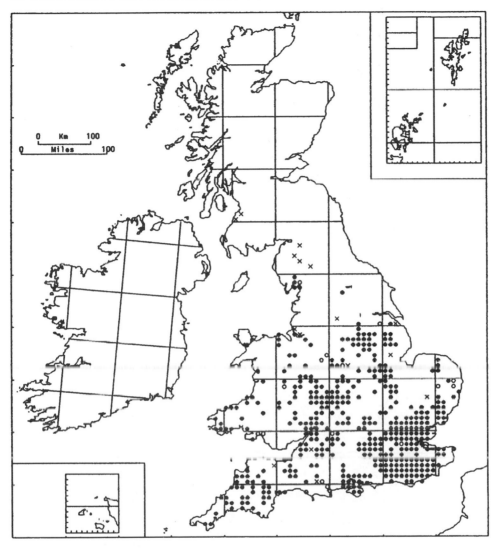

Figure 1.3. Distribution by 10-km squares of the wild service tree *(Sorbus torminalis)* in England and Wales: • 1950 onwards, ○ before 1950, × introductions. Reproduced by permission of the Botanical Society of the British Isles from P. Roper (1993), *Watsonia,* **19**, 209–229

a fashionable tree, part of the Standard Broadleaved Mixture. What good does that do? Although rare, . . . it is not threatened. Its meaning lies in being a rare and wonderful tree with a mysterious natural distribution (Figure 1.2). It is devalued by being made a common tree' (Rackham, 1991). *S. torminalis* shows a modern pattern of distribution which reflects to some extent the part the tree has played in the life of the countryside over many centuries (Figure 1.3; Roper, 1993).

Plate 1.2. Dry forest in Santa Rosa National Park, part of the Guanacaste National Park Conservation Area, Costa Rica. Dry forest is the most endangered habitat type in Mesoamerica, and the 700 km² park will bring back together once fragmented and damaged forest fragments, forming an area large enough to maintain healthy populations of all animals, plants and habitats known to have originally occupied the site (J.W. Spencer, English Nature)

Plate 1.3. Cattle pasture in the Santa Rosa National Park, Costa Rica. Cattle eat the fruits of the guanacaste tree *(Enterolobium cyclocarpum)* and are the major dispersal agents for the seeds. Consequently, these large mammals are important in the early stages of forest invasion of large expanses of pasture that are to be restored to forest (J.W. Spencer, English Nature)

the conservation of biological diversity is extraordinarily important (Wilson, 1993). For perhaps the first time a fundamentally important aspect of forestry besides timber production has become fashionable (Ratcliffe, 1993). This places foresters in an enviable position to lay the foundations for a series of major contributions to many internationally recognised conservation objectives. It also places a huge responsibility on them to take advantage of the situation now while nature conservation, native woodland and the 'near natural' remain in vogue. The reasons for this are quite clear from the lessons of forest history. Biodiversity will undoubtedly remain important, but the only certainty is that the present political enthusiasm for biodiversity will not stay in fashion for long. Given that the only lasting rationale for attempting to recreate biologically meaningful and diverse forest ecosystems is that of biological conservation, we should be taking every advantage of the present fleeting fashion for native woodland to secure large areas of forest, established in such a way, in both the ecological and the legal sense, that they make a lasting contribution to international conservation strategies (Plates 1.2 and 1.3).

REFERENCES

Batisse, M. (1986). Developing and focusing the biosphere reserve concept. *Nature and Resources*, **2** (3).

Buckinghamshire County Council (1991). Black Park Local Nature Reserve Management Plan.

Emery, F. (1974). *The Making of the English Landscape: The Oxfordshire Landscape*. Hodder and Stoughton, London.

Forestry Authority (1994). *Native Pinewoods. Forestry Practice Guide 7: The Management of Semi-natural Woodlands*. The Forestry Authority, Edinburgh.

Joint Committee for the Conservation of British Insects (1986). Insect Re-establishment – A Code of Conservation Practice. *Antennae*, **10**, 13–18.

Maunder, M. (1992). Plant reintroduction: an overview. *Biodiversity and Conservation*, **1**, 51–61.

Payne, J. (1990). *Wild Malaysia*. New Holland, London.

Peterken, G.F. (1987). Natural features in the management of upland conifer forests. *Proceedings of the Royal Society, Edinburgh*, **93**, 223–234.

Peterken, G.F. (1993). *Woodland Conservation and Management*, second edition. Chapman and Hall, London.

Peterken, G.F. and Game, M. (1984). Historical factors affecting the number and distribution of vascular plant species in the woodlands of central Lincolnshire. *Journal of Ecology*, **72**, 155–182.

Peterken, G.F. and Backmeroff, C.E. (1988). *Long-term Monitoring in Unmanaged Woodland Nature Reserves*. Research and Survey in Nature Conservation 9. Nature Conservancy Council, Peterborough.

Pigott, C.D. (1991). *Tilia cordata* Miller: Biological Flora of the British Isles. *Journal of Ecology*, **79**, 1147–1207.

Rackham, O. (1980). *Ancient Woodland*. Edward Arnold, London.

Rackham, O. (1986a). Ancient woodland of England: the woods of south-east Essex. Rochford District Council.

Rackham, O. (1986b). *The History of the Countryside*. Dent, London.

Rackham, O. (1991). Landscape and the conservation of meaning: Reflection Riding Memorial Lecture. *Royal Society of Arts Journal*, **139**, 903–915.

Ratcliffe, P.R. (1993). *Biodiversity in Britain's Forests*. The Forestry Authority, Edinburgh.

Rodwell, J.S. (ed.) (1991). *British Plant Communities*, vol. 1: *Woodlands and Scrub*. Cambridge University Press, Cambridge.

Rodwell, J.S. and Patterson, G.S. (1994). *Creating New Native Woodlands*. Forestry Commission Bulletin 112. HMSO, London.

Roper, P. (1993). The distribution of the wild service tree, *Sorbus torminalis* (L.) Crantz., in the British Isles. *Watsonia*, **19**, 209–229.

SEDUE (1987). *Plan de Manejo de la Reserva de la Biosfera Sian K'aan*. Sedue, Chetumal, Mexico.

Soutar, R.G. and Peterken, G.F. (1989). Regional lists of native trees and shrubs for use in afforestation schemes. *Arboriculture Journal*, **13**, 33–43.

Taylor, C. (1983). *Village and Farmstead*. George Philip, London.

Tubbs, C.R. (1986). *The New Forest*. Collins, London.

Williams, M. (1989). *Americans and Their Forests: A Historical Geography*. Cambridge University Press, Cambridge.
Wilson, E.O. (1993). *The Diversity of Life*. Allen Lane/The Penguin Press, London.

2 Planning and Designing New Woodlands for People

M. SANGSTER

INTRODUCTION

Woodland is probably the most potent means available for bringing about extensive landscape change. It is the only extensive, productive alternative to agriculture. It can provide new habitats of great value to wildlife. Increasingly it features not only in rural planning but in development planning generally.

Of all rural activities in the United Kingdom woodland management is probably the most regulated, mostly through non-statutory voluntary agreements. Anyone planning and designing new woodlands should be aware of the potential constraints which this imposes. There will almost certainly be national and regional guidance on the scale and type of woodland or forestry activity favoured in any locality; and if grant-aid is sought then the granting agency will have particular requirements.

Recent research in the USA (Dwyer and Hutchinson, 1990) and the UK (Burgess, 1993) shows that a person's social and ethnic background greatly influences their attitude to woodlands. So a woodland's location, both geographically and in respect of the people likely to use it, is a most important factor in planning and design. In the UK the Forestry Commission's cost-benefit appraisals of new woods (Forestry Commission, 1993) has shown that the most significant benefits they provide are often recreation and access for local urban populations. Ironically, research carried out for the Countryside Commission highlights how alien an environment woodland can be for our urban population (Burgess, 1993).

This chapter looks first at these wider social and planning issues and then more specifically at people's attitudes to woodland. These are key issues in modern woodland design. However, the starting point for anyone designing woodland is, as it has always been, to establish clear long-term management objectives.

'ECOLOGICAL INFRASTRUCTURE'

The national perspective

In most western countries agriculture and forestry are subsidised by their Governments, often to meet a range of social objectives. Thus in the UK since 1919, when the Forestry

The Ecology of Woodland Creation. Edited by Richard Ferris-Kaan

Commission was established, the objectives of Forestry policy at different times have included:

- maintenance of rural populations
- provision of countryside recreation
- protection of environmental features
- enhancement of natural beauty.

Today forestry is regarded as multifunctional (Secretary of State for Scotland, 1991) and public funding is expected to bring public benefits. In addition to timber production all of these objectives are acceptable for grant-aided woodland planting. Indeed a survey undertaken by the Agricultural Development Advisory Service (ADAS) in 1992 shows that even traditional landowners value their woods more for their amenity than their timber. Table 2.1 shows the responses to a questionnaire in which landowners were asked to name their principal management objective.

Table 2.1 Woodland use by private estates in England and Wales

Principal use	Large estates %	Medium estates %	Small estates %	All estates %
Shade	4.1	2.9	14.0	5.6
Landscape	16.4	14.7	19.0	16.1
Nature	12.3	14.3	24.0	15.6
Sport	23.4	25.7	23.0	24.1
Timber own use	18.5	18.9	16.0	18.6
Timber for sale	23.3	21.0	3.0	17.9
Other use	2.0	2.5	1.0	2.1
Total %	100.0	100.0	100.0	100.0

Units: % of replies.

In many countries, for instance Denmark and the USA, management of state forests for multiple uses is a statutory requirement. Forestry, particularly new planting and felling, is often regulated by governments for social reasons. In Switzerland and Austria, for example, avalanche control and soil stability are important concerns. In Britain regulations originally intended to protect a scarce strategic resource are now used to conserve local amenity and wildlife habitats, and environmental assessment is a requirement in all European Union countries when forestry is likely to result in significant environmental change.

Two examples, from Scotland and southern England, show how forestry planning in the UK is becoming increasingly sophisticated and that the issues in rural and in heavily populated areas are very similar.

The regional perspective

In Scotland the regional authorities worked with the Forestry Commission to develop strategies based on land capability for locating new forests, in which nature conservation was also a main criterion (Scottish Development Department, 1990). Although the

Forestry Commission has well-developed guidelines on forestry and landscape, local authorities wished to include landscape criteria in their plans. A number of local authorities are now undertaking landscape assessments with the aim of developing guidance on the scale and type of forestry in a locality (see Bell, Chapter 3).

However such strategies are evolving still further. In 1991 the Forestry Commission introduced grant-aid specifically to encourage new woodlands for public access in areas of identified need. This provided the local authorities and Forestry Commission with an opportunity to look at where populations are located and the use that people make of their local countryside and draw up plans for the strategic application of the new grant. Effectively, forestry in the Scottish Regions is influenced by land-use and social considerations at three levels of definition:

● Land capability and alternative land uses.
● Local landscape requirements: scale and nature of the woodland.
● Social considerations: where people are located and the use they make of the countryside.

In England Indicative Forestry Strategies (Department of the Environment, 1992) are less developed but another layer of definition is appearing. Land capability is much less of an issue since there are far fewer areas where trees will not grow. However it seems likely that heavily populated local authorities will look at demographic and economic development trends when they develop their forestry strategies. Thus the very densely populated southern English counties of Kent, Hampshire and Berkshire are each concerned to use woodland to ameliorate major road, rail, and housing developments forecast in their development plans (Plate 2.1). Growth in extractive industries and forecast population trends are also a part of this equation.

Plate 2.1 Many townsfolk would perceive this as a 'natural' scene: a diverse botanic garden, within an urban setting (M. Sangster, The Forestry Authority)

However woodland is still not accorded equal importance with other land uses. Two examples of woodland policies in development planning illustrate the relative importance placed on woodland.

1. In 1991 South Hertfordshire County Council attempted to include policies in its draft structure plan so that farmers would be allowed to redevelop farm buildings in return for planting a large part of the farm with trees (Pitt, 1991). This use of planning gain was controversial and the proposals were removed from the plan at the public examination.

2. In 1993 local authorities in the National Forest in the East Midlands included general policies in their development plans that tree planting should be included in significant residential and commercial developments and that the landscape strategy developed by the National Forest should strongly influence the design of the planting (Countryside Commission, 1993).

In this second example there is no planning gain traded; the inclusion of woodland in a development is secondary to the decision on whether the development is approved or not. The author does not know of any cases where local authorities have used their powers under the planning acts specifically to create new woodland. We still have a long way to go before woodland is seen as a local benefit comparable, say, to a new road layout or a car park. However it can be argued very persuasively that significant woodland planting will not take place in the urban fringe, where land values are much higher than in rural areas, unless planning gain is used to finance it (Pitt, 1991).

Despite this there are good examples where woodlands have been used to repair damage to landscapes caused by past industry, farming and land speculation. This is one of the main objectives of the Community Forestry initiative and it is a well-tried approach. The community forestry initiative was launched by the Forestry Commission and Countryside Commission in 1989. Its aim is to improve degraded landscapes and provide for recreation and education in the countryside close to 12 major English towns. Other well-known examples using this approach include the river valleys initiatives in Manchester in the 1980s and in Strathclyde today, and the 'Waste into Woodland' projects by Groundwork Trust in the North-west.

Many local authorities have woodland strategies. Invariably they have an environmental focus. Many are concerned to encourage management of existing woods and most seek to identify priority areas for new woodlands. In England, Leeds, Calderdale and Plymouth are good examples.

WOODLAND BENEFITS

Economic arguments for nature conservation

Two recent pieces of research by the Forestry Commission in the UK (Forestry Commission, 1990; 1993) have attempted to place values on the intangible benefits of new woodlands. The studies highlighted recreation and amenity value to local residents as the main benefits of new woodland.

Research into peoples' preferences for woodland (Lee, 1992) showed that people think nature conservation is a most important benefit of forestry. However, research found little existing applied economic research into nature conservation benefits. If nature conservation is to compete with other land uses for government funding it will be far more successful if it has a sound underpinning in economic theory. This lack of a theoretical base is illustrated in an attempt by Good *et al.* (1991) to value woodland nature conservation, where they resort to looking simply at the additional costs of designing and managing woodland specifically for this objective. Such an approach does not acknowledge that a painting is worth more than the artist's paint and canvas.

Public perceptions of woodland

Research for the Countryside Commission (Burgess, 1993) investigated the attitudes of urban people to woodlands. It showed that many people felt insecure in them. Women in particular felt threatened, worried by the prospect of being attacked. In reality woodlands are an extremely safe environment. However if we wish urban people to use new woodlands we must design them to minimise these perceptions of threat. This research is paralleled by a large body of work in the USA (Dwyer, 1988). A recurrent finding is that different social and ethnic groups perceive woodlands differently. In the context of design there is a strong message here: start with a clear idea of who it is you want to benefit from your new woodland.

These findings have also been reinforced by the work of Lee (1992), which looked at peoples' attitudes to forests and landscape preferences. The study was based on questionnaires in one part of which respondents were asked to rank four benefits of forestry; their ranking was: nature conservation, scenery, recreation and timber production. The main conclusions of the research were:

- Most visits to forests were made by car but the small proportion (19%) of visitors who walked in accounted for a great number of visits (dog walking accounted for 20% of all visits).
- Forest visits are most likely to be by professional and managerial people.
- The most positive reactions to woodlands were spiritual and aesthetic in nature.
- Negative feelings, expressed particularly by women, included fears of vulnerability and getting lost.
- Almost 70% of the people in the survey thought that lay people should have more say in the design of forests.
- In terms of forest landscape preferences diversity was found to be the most important characteristic, followed by water bodies (another element of diversity?) and attractive open spaces.

There is a remarkable similarity between woodland owners and visitors in their views on the benefits of woodland.

DESIGNING FOR PEOPLE: THE SOCIAL ISSUES

In Britain only 2% of the population is employed in agriculture or forestry; 80% of the population lives in towns, a figure which has changed little this century. For most people woodland is an alien environment into which they bring all the preconceptions and anxieties of the town.

Anxiety: perceived threats and the use of open space

In the study undertaken by Burgess (1993) townspeople were taken into woods and an attempt was made to analyse their responses. She showed that women can often feel threatened in woodland and are worried, perhaps quite unconsciously, that an attacker might be hidden by the trees or undergrowth. However middle-aged and older women were more confident. A group of Asian women with whom she worked were particularly unsettled in the woodland environment. Men also felt uncomfortable. Getting lost and trespassing, with the possibility of confrontation with 'Authority' was one of their concerns.

Householders are another group who may be concerned about woodlands. Thus in Edinburgh in Scotland in 1994 a proposal to establish community woodlands adjacent to a housing development was opposed by the residents on the grounds that the trees might hide people of criminal intent. Possible attacks on children were cited as a major concern. In Redditch in central England, a new town with many areas of communal woodland, the local authority receives frequent complaints from householders worried that adjacent woodland provides cover for burglars.

This last concern is the only one that seems to have some grounds. Research of criminal statistics showed that otherwise woodlands are extremely safe places and the chances of attack or other abuse are much less than in, say, a typical town park in which all the groups would have felt at home (Burgess, 1993).

Woodland design will not prevent such anxieties. However imaginative use of open space in and around woodlands might alleviate them. Techniques include:

- *Sightlines:* ensure that paths have good longitudinal visibility, avoiding sharp bends.
- *Margins:* paths should have open space to either side. Where the vegetation comes up to a path it should be opened up to allow visibility through it. Open verges also are often recommended for nature conservation.
- *Path widths:* they should be wide enough or have passing places so that people can pass comfortably.
- *Lighting:* in some circumstances lighting will be appropriate.
- *Thinning:* Forestry Commission staff will often differentially thin woodlands with low stem densities and good visibility into the crop close to paths and roads.
- *Undergrowth:* bushes and tall herbaceous vegetation are viewed with great suspicion by many townspeople. Mowed verges and low vegetation are preferred.

However to some visitors this might be unwelcome suburbanisation of the countryside. The approach adopted by Forestry Commission landscape architects (Forestry Commission, 1991) is to offer visitors a choice of experience by using these techniques but also having areas which are much more natural. A superb example from continental Europe is the Amsterdamse Bos in the Netherlands.

The needs of different social groups

Specific social groups have particular needs. If a woodland is being designed for recreational use then it will be necessary to do some market research to find out who is most likely to use it and what particular needs they may have. For example, if a woodland is close to a town and accessible on foot then it is more likely to be used by different groups to one which is accessible only by car; perhaps by young mothers, elderly and unemployed people who are out and about during the day. Woods accessible by car will be busier at weekends and in the evenings.

Religious and ethnic groups also have their own preferences. Minority groups in the UK make up a disproportionately small fraction of woodland users. The reasons are unclear though British Trust for Conservation Volunteers in the West Midlands had good success in encouraging Asian women's groups to take part in environmental action. Their success stems from the effort taken to tailor their approach to the cultural mores of the community.

Woodlands close to schools are likely to be used for educating the children (Plate 2.2). People from old folk's homes, local office blocks, nearby car parks, shopping centres, athletics clubs, scout groups and others are likely to make use of accessible woodland for their own purposes. In attempting to get an idea of the scale and type of demand the questions to ask are:

1. Who is going to use the wood? *Location* is the main determining factor.
2. What do local people want from a new wood? (If they want one at all).
3. What are the implications for existing uses of the site?
4. What scope is there for community involvement in planning, planting and caring for the wood?

Plate 2.2 Getting local people to participate: children with a planting bag on a Community Forest site (M. Sangster, The Forestry Authority)

Criminal damage

In a survey by the National Forest (Countryside Commission, 1992) landowners in the project area were deeply concerned by proposals that new woods should be opened to the public. Fear of criminal damage was the most common reason given. In urban areas woodland designers also seem excessively concerned by this threat. There is good empirical evidence that in practice new woods are rarely damaged and when they are it is often predictable: usually they interfere with some other use of the site that the local community values. The following are two examples:

- Poplar trees planted in 1985 in Stockport were cut down by adjoining householders because they threatened to obscure cherished views of the Pennines.
- Fences and trees on landscaped coal tips in 1982 in Newcastle were damaged by off-road motor-cyclists who traditionally used the tips and had no alternative practice areas.

Again these difficulties might not have arisen if some market research had been done before the trees were planted.

It is possible to reduce criminal damage through good design. The following advice is from the Black Country Urban Forestry Unit in the West Midlands:

- Small planting stock appears to attract less attention than whips and standards.
- Good paths reduce the motivation to wander through a planting scheme.
- Denser planting for 5 metres or so along paths and edges can reduce ingress, particularly if spiny species are used.
- Fly tipping can be reduced by ensuring that car access is restricted to a small number of points.

Community development

The Forestry Commission's approach increasingly is to recommend that local environmental and community groups should be used to do this initial market research. We believe that this will also encourage greater use of the wood and reduce casual vandalism. Dealing with communities is a skill in its own right and these groups will have a great deal of local experience. In a recent report to the Forestry Authority, Ageyman (1993) recommended a six-point plan for involving communities:

1. Appoint an experienced 'community co-ordinator' to research local views and to negotiate access to institutions.
2. Find out who is going to use the wood.
3. Arrange local meetings to open consultation before the planning begins.
4. Develop links with useful organisations and individuals (network).
5. Provide appropriate training for the people who are going to help create the wood.
6. Draw up an action plan to deal with organisation, evaluation and follow-up.

Obviously this is an approach suited to a large-scale initiative but elements of this procedure will apply to small schemes.

CONCLUSION

Since so many people live in towns it makes sense to concentrate woodlands in places where they are accessible, or at least visible, to townspeople. A number of more or less experimental initiatives in the UK have been established in the past few years to promote forestry in and around towns and in populated areas. They include The Community Forestry Initiative, The National Forest and its Scottish sister, the Central Scotland Woodlands Initiative. The Forest of Belfast and Black Country Urban Forestry Unit are urban initiatives aiming to encourage forestry within towns. The Welsh Valleys Initiative aims to redesign Forestry Commission woodlands to meet the wishes of local people and the Sherwood Initiative uses woodlands to repair damaged landscapes in the East Midlands.

Sustainability and biodiversity are the new environmental concerns and of course human beings have similar environmental needs to other mammals. Woodlands offer a most potent means for applying these concepts to the human, largely urban, environment.

REFERENCES

ADAS (1992). *The British Market for Forest Thinnings*. ADAS Farm Building Development Centre Internal Publication.

Ageyman, J. (1993). *Guidance on Promoting Community Participation in Urban Forestry Activities*. A report to the Forestry Commission. Forestry Commission, Edinburgh.

Burgess, J. (1993). *Perceptions of Risk in Recreational Woodlands in the Urban Fringe*. A report for the Community Forestry Unit, Countryside Commission. Countryside Commission, Cheltenham.

Countryside Commission (1992). *A Survey of Landowners Attitudes to Public Access in the National Forest*. Countryside Commission, Cheltenham.

Countryside Commission (1993). Appendix 7. In: *Draft National Forest Plan*. Countryside Commission, Cheltenham.

Department of the Environment (1992). *Indicative Forestry Strategies*. Circular 29/92. HMSO, London.

Dwyer, J.F. (1988). Predicting daily use of urban forest recreation sites. *Landscape and Urban Planning*, **15**, 127–138.

Dwyer, J.F. and Hutchinson, R. (1990). Outdoor recreation participation and preferences by black and white Chicago households. In: Vining, J. (ed.). *Social Sciences and Natural Resources Recreation Management*. Westview Press, Boulder, Colorado, 49–67.

Forestry Authority (1991). *The Community Woodland Supplement. Woodland Grant Scheme. Applicants Pack*. Forestry Commission, Edinburgh.

Forestry Commission (1990). Cost Benefit Assessment Undertaken for the Community Woodland Supplement. Unpublished internal report. Forestry Commission, Edinburgh.

Forestry Commission (1991). *Community Woodland Design Guidelines*. HMSO, London.

Forestry Commission (1993). The Costs and Benefits of Planting Three Community Forests. Forestry Commission, Edinburgh.

Good, J., Newton, J. *et al.* (1991). *Forestry Expansion: A Study of Technical, Economic and Social Factors. Forests as Wildlife Habitat*. Forestry Commission Occasional Paper 40. The Forestry Commission, Edinburgh.

Lee, T.R. (1992). *Attitudes and Preferences for Forest Landscapes*. A report to the Forestry Commission and Countryside Commission for Scotland. Forestry Commission, Edinburgh.

Pitt, J. (1991). Community Forest Cities of Tomorrow. *Town and Country Planning*, **60**, 188–190.

Secretary of State for Scotland (1991). *Forestry Policy for Great Britain*. HMSO, London.

Scottish Development Department (1990). *Indicative Forestry Strategies*. Circular 13/1990. HMSO, London.

3 New Woodlands in the Landscape

S. BELL

INTRODUCTION

In most western countries we live in a landscape which is far from natural. Many places are almost entirely the product of human intervention. City landscapes are the most obvious examples. On a grander scale, whole landscapes can be human creations such as reclamations from the sea, especially in the Netherlands. On a subtle level, many places which seem wild to us today are either in highly altered semi-natural states following earlier human abandonment, or are recolonisations by nature of former inhabited areas (Hoskins, 1955). Some are even deliberate restorations of landscapes which suffered some form of catastrophe in the quite recent past, well illustrated by the three examples that follow. The rest of the chapter explores something of the evolutionary character of woodlands in the landscape and suggests ways in which to create new woodland in harmony with an understanding of this.

- *Dartmoor*
  On Dartmoor in Devon, in the south-west peninsula of England, the scenically highly attractive moor with heather (*Calluna vulgaris*) and other semi-natural vegetation was once wooded. It was later cleared for agriculture during the neolithic period (2500 BC). Over the next 2 millennia, the soil suffered exhaustion, the climate became cooler and wetter and the area was gradually abandoned. Due to podzolisation, the soil was too poor and infertile to support anything but acid-loving heathy plants which in turn support a few sheep (Rackham, 1986). The resulting 'wild' landscape is now a National Park, designated for its 'natural beauty' (Countryside Commission, 1993).
- *Eastern USA*
  In the eastern states of New England in the USA, colonists in the 17th and 18th centuries cleared much of the natural woodland for agriculture. After a few decades many of the more marginal areas lost fertility and farmers often abandoned the land to move west to areas which promised better farming. Their former farms reverted to second-growth forest (Pyne, 1982).

The Ecology of Woodland Creation. Edited by Richard Ferris-Kaan.

● *South-eastern USA*
In the southern Appalachians, restoration of the landscape was carried out in the 1930s. The people of the area became poverty stricken when their economy collapsed. This was in part due to the Depression, but part through the effects of fire and chestnut blight which killed trees that were economically important to the people (Pyne, 1982). Under the New Deal, the Government set about reforesting many areas. Today, the Shenandoah National Park occupies some of this land which to the casual visitor looks as if it was always forest (Wilson, 1992).

Patterns of woodland change

This kind of pattern, of colonisation and clearance of woodland and forest followed by some form of degradation and later by re-establishment of woodland, has recurred throughout history and could well be an important theme in the next few decades (Hoskins, 1955; Rackham, 1986). So much land throughout the world, in areas where forest would naturally be the climax form of vegetation, is degraded or exhausted in some way or other. This may have resulted from inappropriate forms of husbandry, from population pressure or from industrial dereliction. The response has often been to plant up areas, perhaps with non-native trees, in the form of single age, single species plantations. Britain's reafforestation programme concentrated for many years on conifers in plantations in order to re-establish a timber supply. In other places, eucalypts (*Eucalyptus* spp.) or other fast growing species have been extensively used. While there were good reasons for choosing such species at the time, and while it is possible to improve the way such areas are designed and managed, with hindsight it can be seen that other, more subtle approaches could have been adopted. Methods are needed which avoid over-technological approaches, where sites are altered to suit the species and management techniques, and instead work more harmoniously with the landscape setting and with the dynamics which continue to affect the evolution of the landscape. Action taken today will in the future appear to be just one more stage in the ebb-and-flow of change.

Decisions about where and what to plant have often been taken at the site level, with very little attention paid to the larger, 'landscape' scale factors. In order to take these into account the following information is needed:

● A knowledge of how and why a particular landscape has evolved and changed.
● Broad criteria for the location, scale, position, type and design of new woodlands at the 'landscape' level.
● Design principles to apply to woodlands at the site scale.

WOODLANDS IN AN EVOLVING LANDSCAPE

In many ways, the story of pre-industrial, pre-urban landscape is one of progressive deforestation. Man arrived, population expanded and agriculture developed and spread. This resulted in a number of distinct landscape types often overlaid on one another: *ancient, feudal, planned and designed, industrial and bureaucratic*. In Europe, this story has taken a long time and is by no means one of continuous advance by agrarian settlement. In North America and in many other recently colonised countries, technology allowed the progress of centuries to be accomplished in a few decades.

When designing new woodlands in the landscape it is important to know what the patterns and dynamics are in a particular area (see Spencer, Chapter 1). What have been the influences in a particular place and how can we take them into account? We need broad criteria for the location scale, position, type and design of new woodlands in the landscape. This can only be achieved by some sort of landscape planning and design.

Arrival and expansion of human settlement

In Britain, serious clearance for agriculture started in around 4000 BC at the commencement of the neolithic period (Rackham, 1986). It is probable that many clearings were recolonised by woodland so that the landscape changed only sporadically. Areas were probably chosen at first because they could be cleared easily and cultivated using primitive tools (Hoskins, 1955). There is evidence that some woodland areas were actively managed at this time, probably by coppicing, in order to produce large supplies of small diameter wood for hurdles, palisades, huts and causeways (Rackham, 1986). With some variation depending on tribal practices, the customs of more recently arrived immigrants and direction by some higher authority in certain places (for example, Wessex at the time of the great constructions at Stonehenge in Wiltshire (Clarke *et al.*, 1985) or the time of the Roman occupation), this process continued.

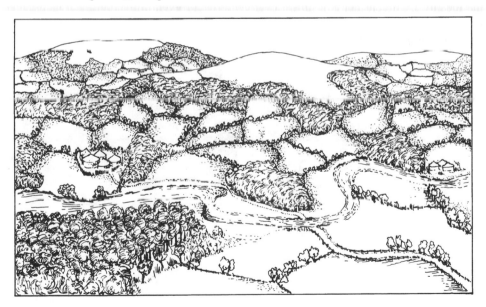

Figure 3.1. An ancient British landscape as it might have looked when people were making the original clearances from the forest. Fields are pushing outwards and grazing is expanding down from the higher ground

This first phase is typified by small settlements, perhaps by families or family groups, establishing themselves in an area and clearing the land (Figure 3.1). They may even have claimed some sort of primitive form of ownership over it by erecting a family burial mound or other method of site marker where oral traditions and family stories could legitimise the rights of occupancy (Clarke *et al.*, 1985). At certain periods it is possible that some groups were pushed up into agriculturally marginal areas by local population pressures. If a harvest failed, they were vulnerable to famine so the land might quite easily become abandoned and the woodland could have reasserted itself quite readily. There was therefore a kind of frontier in places, which fluctuated over time (Hoskins, 1955).

A second phase of consolidation then followed: as populations expanded and organised government developed a system of land tenure, land laws and gradual domination by the rich and powerful occurred. Highly developed land tenure and social organisations became established.

Ancient landscapes

The landscape pattern which developed over these two phases is surprisingly intact in many parts of Britain. It is a pattern often referred to as 'Ancient' (Rackham, 1986). The key feature of this is that land use developed as a gradual process carried out simultaneously by numbers of individuals or small groups without any overall guide or direction. Thus this landscape is 'self-organised' with the patterns often showing great degrees of similarity and cohesiveness over great distances. They demonstrate a 'goodness-of-fit' to the locality from constant adjustment over centuries or even millennia. Such landscapes are not static: they have altered at their margins and some have been abandoned for periods, for example, following the Black Death.

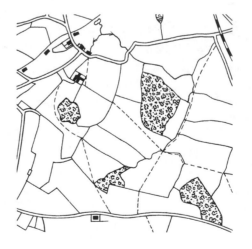

Figure 3.2. A plan of a typical 'Ancient' landscape, showing winding lanes, irregular fields, ancient woods and nucleated settlements

The ancient landscapes (Figure 3.2) are characterised by irregular sized and shaped fields with curving boundaries, winding lanes, nucleated settlements and ancient woodlands set among the fields (Rackham, 1986). Earth banks, ancient hedges and walls, sunken lanes, and other similar features are still in use, yet date from long ago. In the uplands it may be possible to see where the fields formerly extended up into areas now heath or moor, such as the Dartmoor reaves (Fleming, 1988).

Feudal landscapes

In the Midlands and south of England in particular, the great open field systems were developed in the Middle Ages. These were open landscapes of cultivated fields organised in the strip system, with common grazing on heath, open pasture and in nearby woodlands (Steane, 1985). The woodlands might often have been open wood pastures while coppice woods would have been fenced or hedged to keep out grazing and browsing animals (Hoskins, 1955; Rackham, 1986).

There was also an increase in woodland cover as part of the establishment of hunting forests including heath, bog, open water and scrub. Quite significant areas were created during the early medieval period, with their own laws and officers (Harrison, 1992). Farms and villages were regularly cleared away in order to establish the forests.

In the 16th century, woodland constituted a generally diminishing component of the landscape (although locally increasing), existing as managed coppice, wood pasture and park and as a component of the hunting forests. At the edges of cultivation the extent of woodland fluctuated (Hoskins, 1955). In the 17th and 18th centuries, the landscape in many places was to change dramatically.

Planned and designed landscapes

These landscapes are largely, but not entirely, the result of the Parliamentary Enclosure Acts of 1720–1840 (Rackham, 1986). Enclosure of the great open field systems started in many places after the decrease in population by the Black Death (Hoskins, 1955). This reduced the labour force which led landowners to favour less labour intensive animal husbandry over arable cultivation. These enclosures produced landscapes different from the 'ancient enclosures' made directly from forest or moor. They were less irregular with straighter lines and many often comprised subdivision and simplification of the old strip system of tenure (Figure 3.3).

By the 18th century the agricultural revolution had begun. Large landowners began to enclose open fields, commons, heath, old hunting forests and parks by obtaining an Act of Parliament which allowed them to override and extinguish the rights of common over many of the areas (Hoskins, 1955; Rackham, 1986). Surveyors laid out completely new landscapes with a few former features, such as parish or estate boundaries or roads. New farms were built and the new fields enclosed with hedges of quick-thorn and spaced trees or walls. Very few open field systems survived. Planned landscapes may superficially resemble the other enclosed landscapes, but they are quite different. They were created over large areas by a single event under the control of the landlord. They also relied more on advances in drainage, crop rotation, lime application, shelter plantings and other agricultural developments to bring out their fertility, thus beginning to break away the dependence on microsite and microclimate of previous generations of farmers.

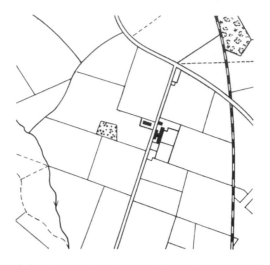

Figure 3.3. A planned landscape, showing straight roads, rectangular fields and plantation woodlands

Designed landscapes, the great parks and pleasure gardens attached to the great country houses, came about as earlier farms, parks and gardens were swept away. Where enclosure by woodland and control of vistas by carefully positioned clumps and groups of trees were central to the designed effect, new landscapes were created (Hussey, 1967).

Industrial and bureaucratic landscapes

With industrialisation the agents of change were less and less concerned with the soil, local landforms and climate, as technology was able to overcome them. Rural landscapes changed in two ways. Firstly, industrialisation and imperial expansion, together with falling grain prices after the repeal of the corn laws, meant that more and more food could be imported. Cereal production was curtailed, fields were taken over by rushes (*Juncus* spp.), bracken (*Pteridium aquilinum*) spread and woodlands advanced in some places. Then during the two World Wars, in particular World War Two, food production became a strategic goal (Blunden and Curry, 1988). Since then central government policy, and later the European Community policy dictated down to farmers and landowners through education, advice, grants and subsidies.

In the rural heartlands, hedges, trees and woods were ripped out and continuous arable production replaced crop rotations. Many areas of open downland, upland fringes, former heathland and parkland became arable producing areas. In the uplands, moorland was 'reclaimed' to pasture or afforested with conifers. Shelterbelts and drainage helped the increase in sheep numbers which reached record proportions in the late 1980s. Certain landscapes were protected as National Parks, National Scenic Areas or Areas of Outstanding Natural Beauty.

The collapse of industry coupled with urban expansion, and the development of transport corridors conspired to produce a jumble of ill-associated elements within the landscape, e.g. in the areas between nearby towns or cities or amongst the greater metropolises.

Woodland and trees in the landscape

In all of the landscapes described above, woodland and trees have been, remain or could be an important component. In fact, in the ancient landscapes carved from the original wild-wood, the ancient semi-natural woodlands and hedgerows provide structure and continuity with the past. The Weald of Kent still appears more wooded than open in many parts. In the upland fringes of Wales or Cumbria, valley woodlands and areas upon the hillsides are testament to the continuing ebb and flow of agricultural colonisation at its most tenuous. Merely reducing sheep numbers and abandoning fields would encourage this woodland to reassert itself as has already been seen in some areas of heathland (Rackham, 1986).

It is the almost complete loss of woodland and tree cover in some of the Midlands, East Anglia and the southern counties of England which has resulted in the devaluation of their landscape quality. As the older character has disintegrated, the sense of place has disappeared, the richness of the fabric of the landscape has become progressively more threadbare and the sense of continuity with the past has been broken.

In other countries, some of the same influences have been at work. Colonisation phases can be observed in countries such as the USA or Canada, starting in the East and moving westwards, sometimes jumping unattractive areas such as dry prairies and return-ing to them later on. Population pressure, technology such as steel ploughs and railroads helped speed up the process. However, families on marginal land such as parts of Oklahoma or South Dakota fell victim to crop failures and drought, as recently as the Dustbowl period of the 1920s and 1930s (Brogan, 1985). In Finland, for example, primi-tive slash-and-burn agriculture was practised until late last century in places, and the forest has re-established itself since it was abandoned (Antikainen, 1993).

The planning phase was important in the USA where huge areas of the mid-west were laid out in vast rectilinear grids by surveyors. Land was given or sold off according to this grid, roads and settlements established and local government boundaries defined. In Ireland, the Anglo-Irish ascendancy in the 18th and 19th century led to large areas being laid out as regular field patterns in the more fertile east and south. Demesnes (or parks) around country houses were also established and often contained the most significant woodlands in the area. Following independence from Britain, the big estates were broken up into smaller farms. The small field sizes and the density of hedges make the field patterns some of the most intense anywhere.

Abandonment of the land has also occurred and is still happening. In Ireland, in parts of the midlands and west, land is vacant, fields are uncultivated or ungrazed, the hedges are growing out and scrub is reinvading. Woodland will soon take over.

LANDSCAPE PLANNING FOR NEW WOODLANDS

From the resumé of the history of the landscape it can be seen that there are many areas of very different character. The geology, landform and climate varies, and with it types of agriculture, building materials and styles, the original native vegetation and so on. There is concern among the many people who care for the landscape — whether as scenery, for its historical connections or its wildlife values — that its quality will continue to decline (Crowe and Mitchell, 1988). Attention has focused on arresting this, but is now shifting to more positive measures, i.e. restoration, recreation or creating new landscapes (Forestry Commission, 1991). It is important that the varying character of the landscape is fully taken into account, so as to reflect both the inherent richness found there and to maintain the regional and local distinctiveness which combine to provide a sense of place (Bell, 1993).

Landscape character can be defined on a variety of levels, from a continental and national scale down to regions and much more local areas, parishes perhaps, where people have a sense of belonging. The appropriate level at which to invoke this character depends on what level the planning is to take place. For most practical purposes, there would seem to be two useful levels, the regional and local which, in combination, may be used to formulate guidelines.

Regional planning

A regional assessment of landscape character can be used for more general guidelines on where and what kind of woodland might be appropriate without becoming too obsessed with detail. A broad picture is needed, guided by the broader influence in the landscape. Such an assessment can be carried out by computer analysis of different factors described for kilometre squares or some similar method (Countryside Commission, 1993). The factors which might have influenced the development of the landscape pattern can be keyed out for sample squares. They might include elevation, slope, geology (solid and drift), soil, natural vegetation, crop types or husbandry, tree cover, settlement and building types and climatic information, among others. Analysis techniques such as TWINSPAN (Two-Way Indicator Species Analysis; Hill, 1979) can test the relationships between these factors, and from this a map of relatively homogeneous areas can be produced. These not only show up the broad pattern, but relate it to the most important factors which have contributed to its development. In some areas it might be climate, soil and geology; in others, settlement and cultivation or industrialisation (Figure 3.4).

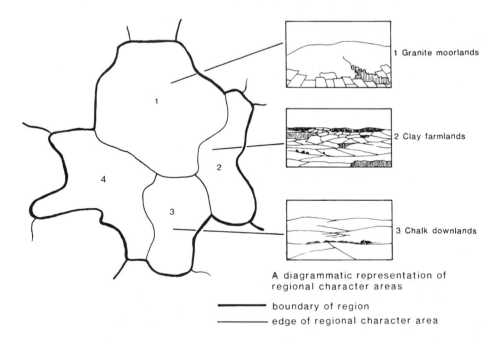

Figure 3.4. A diagrammatic representation of regional character areas

One prototype of such a map has been compiled by the Countryside Commission for England in their 'New Map of England' project (Countryside Commission, 1993). The map of the south-west peninsula brings out distinct patterns which can be perceived on the ground. From it some general guidance can be developed which might include the kinds of woodland, its typical location, the kind of proportions which might be appropriate and so on. This might be of use for policy planning purposes by the Forestry Authority or other national or regional level organisations.

Local planning

Each regional type might be composed of three or four local character types which might be used to define further planning guidelines. These could be different parts of a range of hills, for example, plateaux, escarpments and valleys, or be more concerned with human influences such as the field patterns. Such local character assessments are already being compiled at the local authority level. There are several ways of preparing them. The general approach is to study maps and aerial photographs and check out the initial sub-division in the field where visual descriptions, photographs and sketches would be made (Price, 1993).

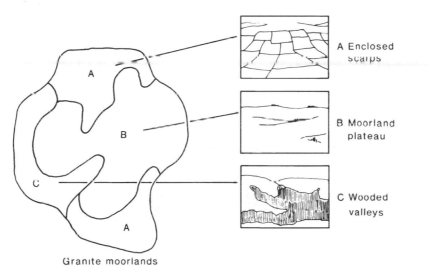

Figure 3.5. One of the regional character areas (as defined in Figure 3.4), broken down into landscape types

The description of each landscape type (Figure 3.5) will reflect the current appearance of the landscape and can be fed back into the knowledge of the key factors which led to its development in the first place, and to the influences causing change. As well as the basic description, such assessments can be used as a base for more detailed design guidance. This may take the form of proportions and types of woodland, together with principles on how to design the shapes and patterns which such woodland will produce. At their most sophisticated they might be used to work with the dynamics of the landscape to help decide where the pattern might 'grow' and develop in a natural way (see Rodwell and Patterson, Chapter 5).

In addition to the description and guidelines, a link might be made with the local woodland types which might arise using the National Vegetation Classification (NVC) (Rodwell and Patterson, 1994; Rodwell and Patterson, Chapter 5). Where new native woodland is desired then a subtle blend could be developed between the natural patterns which this might engender and the cultural patterns which the landscape character assessment has established. This, perhaps mixed with a landscape ecological assessment (in terms of how the landscape functions as an eco-system: see Spellerberg, Chapter 4), would be a powerful tool for getting the best local solutions which answer all the concerns of scenic qualities, cultural and ecological values.

At its heart, however, such planning guidance must not be overly prescriptive since in reality decisions are taken by individual landowners, even if they are steered to some degree by advice and grants. The assessments described above are currently used as an important stage in the preparation of Indicative Forestry Strategies (IFS) (Department of Environment, 1992) which seek to guide landowners and planning agencies on where more woodland is to be preferred. Areas with a recent loss of woodland, with derelict land and large populations might be deemed to be areas where greatest public benefit is to be gained. In other areas there may be sufficient woodland already (intact ancient landscapes perhaps).

Design guidelines

Design guidance has to be developed so that from an understanding of the landscape itself, its patterns, evolution and dynamics, new woodland can be fitted which seems to belong there (Figure 3.6). This is not as simple as might first appear. The objectives for the woodland which the owner of the land may have in mind, the landscape itself and some practicalities which may affect what can be done have to be blended together. This takes some skill and imagination. This involves more than repair and restoration; the possibilities include significant extension of existing woodland or the creation of whole new landscapes in the most devastated or derelict places.

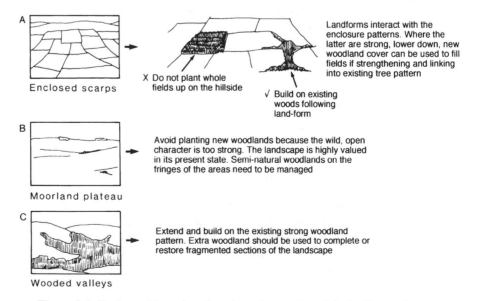

Figure 3.6. Design guidance based on the understanding of the landscape character

PRINCIPLES OF WOODLAND DESIGN

Having some idea of the general type of woodland and its appropriate location in a particular landscape, we can now consider the precise design of some proposed woodland.
Over the years in which woodland design techniques have been developed, some key
principles have repeatedly been found to be useful. These are:

- Shape in relation to landform, vegetation and enclosure patterns
- Scale
- Unity
- Diversity
- *Genius loci* or spirit of place (Forestry Commission, 1989, 1991, 1992; Lucas, 1991;
  Bell, 1993).

Shape

Shape can be determined by landform, by natural vegetation distributions or by the cultivation and enclosure patterns developed by land use. Landform determines shape in two
ways. Firstly, there is an aesthetic response where it has been shown that the eye is led
around landform by moving up valleys and concavities and down ridges and spurs
(known as 'visual forces' in landform). Woodland shapes which respond by following
this, for example, extending up a valley from a lower slope, seem to fit into the landform
better (Figure 3.7). The kind of shape (spiky in jagged, rocky landforms or curving in
rounded ones) is also important. Secondly, natural vegetation patterns are partly determined in their distribution by landform, since this determines shelter, insolation, water
and nutrient movement and so on. Natural wooded landscapes often show a close relationship between species distribution and landform much as the visual force pattern
described above. The NVC can be used to help predict some of these likely patterns
(Rodwell and Patterson, 1994).

Other vegetation patterns may help to determine where new woodland should go and
how it should be shaped. Bracken (*Pteridium aquilinum*) patterns often reflect soil type,
while rushes (*Juncus* spp.) may reflect poor drainage. In flatter landscapes with less obvious landform, the vegetation may be a more useful guide.

Enclosure patterns — hedges, hedgerow trees and relict woods — may be much more
dominant in many places. They may override the landform in importance in some areas.
The distinction between ancient or planned landscapes may lead to very different designs.
Interlocking, branching and irregularly shaped woodlands will belong more comfortably
in the ancient field patterns (Figure 3.8(a)) while more rectangular, regular shapes may be
more acceptable in strongly planned landscapes (Figure 3.8(b)).

Scale

Scale relates to how we perceive our surroundings when compared with our own size and
the amount of landscape we can see. A landscape is large-scale when it presents comparatively wide horizons or towers above us. Scale can be affected by the degree of enclosure which determines the relative sizes of spaces. Trees, hedges and woods can create
strong spatial enclosure, forming a series of 'rooms' in the landscape that can seem quite

Visual forces: the eye is led down ridges and up hollows

An option where the riverside woodland is extended
up into the hollows

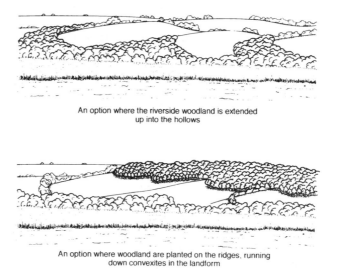

An option where woodland are planted on the ridges, running
down convexites in the landform

Figure 3.7. Landform analysis used to determine when woodland might be fitted into the landscape

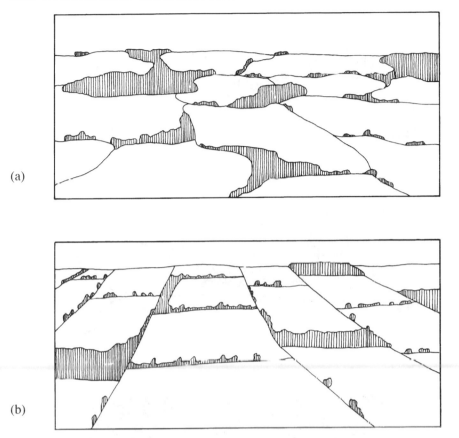

(a)

(b)

Figure 3.8. (a) Organic shapes — irregular, interlocked and branching — fit into ancient land-scapes. (b) Rectangular shapes may be more fitting in strongly enclosed planned landscapes

small. Within a woodland the trees provide even stronger enclosure and yet smaller scale. The perception of scale depends also on the position of the observer. If high on a hill with a panoramic vista, the landscape will appear larger in scale, whereas down in a valley enclosed by slopes and trees, scale will be smaller. As we move through the landscape our perception changes. This should be reflected in design.

In general the average scale of a landscape is determined by a combination of land-form structure and enclosing elements such as trees and hedges. New woodlands will tend to appear most comfortable if they reflect this. In a large-scale landscape, small isolated clumps may look out of place (Figure 3.9(a)) whereas a larger mass of woodland may be much more harmonious. If the landscape appears roughly one-third wooded or one-third open from most viewpoints, this is a good guide as to whether the proportions are balanced (Figures 3.9(b), (c), (d)).

Designs usually work best if they reflect the largest scale of the landscape when seen from the most elevated viewpoint, yet also reflect the smallest scale in the detail of their edges and internal open spaces. There is a hierarchy which should be followed from the larger scale progressively down to the smaller scale.

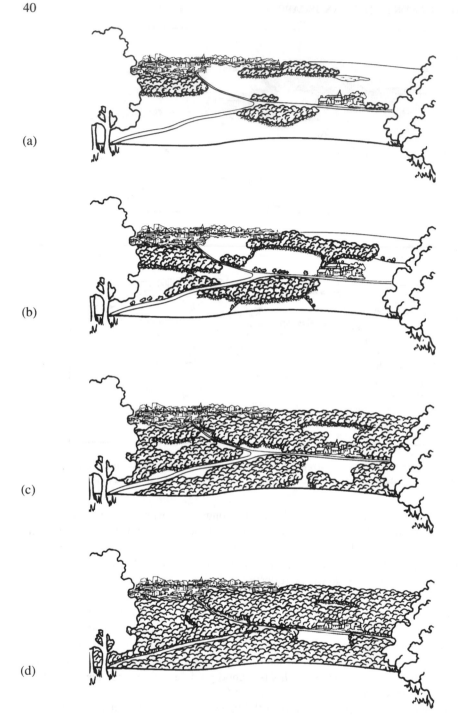

Figure 3.9. (a) Less than one-third of the area is wooded: the woodland 'floats' and is not tied into the landscape well. (b) Around one-third of the landscape is wooded, two-thirds open: this produces a comfortable balance. (c) Two-thirds of the area as woodland allows a good balance of open space to be retained. (d) More than two-thirds of the area is wooded: it looks overpowering and too dominated by the woodland

Unity

Unity means that all the parts of the landscape work together to produce a harmonious whole. If an element stands out perhaps because of an incompatible shape, scale, colour or texture then unity will decrease (Figure 3.10(a)). One of the characteristics of many earlier landscape patterns is their inherent unity: there seems to be a harmony between the various shapes of fields and settlements and the way in which these fit amongst the landform (Figure 3.10(b)). It is often more recent intrusions such as modern buildings or major roads which reduce this sense of unity. Woodland can often be used as an agent for recreating unity by tying such elements back into the landscape (Figure 3.10(c) and Plate 3.1).

Diversity

The richness or variety in a landscape gives immense aesthetic satisfaction. There is a diverse range of landscape types across Britain, Europe and many other parts of the world. Within a given landscape, the different elements provide interest and variety. Too

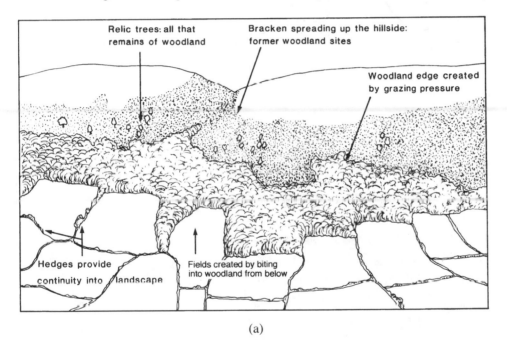

(a)

Figure 3.10. (a) Analysis of a landscape where woodland persists between cleared fields and grazed moors. It could spread and reassert itself if grazing pressure was reduced, fences were improved or fields taken out of pasture. (b) (*overleaf*) Woodland is extended up the hill onto former woodland areas. Shapes are organic and follow landform. The leading edge should be diffused, unless a fence is re-erected and grazing effectively stops further colonisation. (c) (*overleaf*) Woodland is extended down the hill, either occupying whole fields or developing as protrusions or peninsulas from the main body, to create a more interlocking result

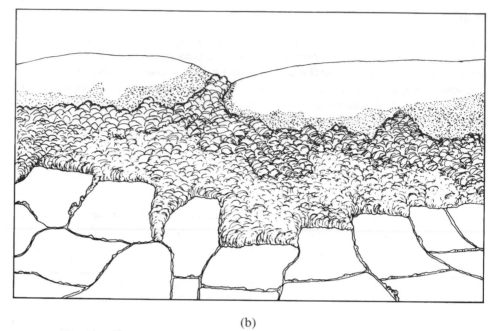

(b)

Figure 3.10b *(continued)*

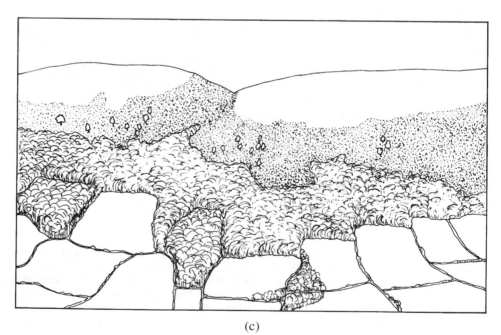

(c)

Figure 3.10c *(continued)*

Plate 3.1. The pattern of fields, farms, woodland and moorland all occupy their appropriate places. The continuity of its pattern, provided by the hedgerows, trees and woods, ties and unifies the landscape together (S. Bell, The Forestry Authority)

much diversity can cause the unity to decline and chaos to ensue: one of the problems of *ad hoc* landscapes. Visual diversity may not always be connected directly with ecological diversity, for example, natural forest may be visually quite boring at a large scale. The two may be more closely related at certain levels of scale, such as different types of native vegetation, rock, soil, water and so on. Man-made elements may contribute visual diversity at the expense of nature, for example, buildings, exotic species, agricultural crops (Plate 3.2). Woodlands can be used to add diversity in appropriate places, or to help

Plate 3.2. A diverse landscape, where the landform, land uses, trees and woodland all contribute to the richness of the scene (S. Bell, The Forestry Authority)

reduce visual chaos by linking different elements or enclosing them. This may provide connecting corridors or networks and greater amounts of edge which are especially important for ecological reasons.

Genius loci or the spirit of place

Each locality is, or was, different from any other and has a particular sense of place or identity (Bell, 1993; Norberg-Schulz, 1980). Sometimes this is generated forcefully by the natural features of a place: water, waterfalls, landform, rare or unusual plants or old trees, for example. It may also arise from ancient human remains such as stone circles, burial mounds, settlements or designed landscapes. The precise qualities which contribute to *genius loci* may be hard to identify exactly, and hence vulnerable to careless action or neglect. Ideally the *genius loci* should be enhanced or recreated as part of the design. This might be achieved by the use of particular localised woodland types or species or by allowing older trees to remain during management, such as the 'granny' oaks of Sherwood Forest (Plate 3.3).

Plate 3.3. 'Granny' oaks in Sherwood Forest provide a strong sense of place, strengthened by all the historical and legendary associations of Robin Hood (S. Bell, The Forestry Authority)

THE PROCESS OF DESIGN

Design starts with an existing landscape which forms the canvas for the designer and a set of objectives which it is desired to meet. The designer's job is to develop a spatial pattern of new woodlands which meet the objectives in a practical and probably economic way while fitting the landscape. In other words, the designer must combine beauty with utility and efficiency.

The objective for the design may be given by the owner and be constrained by national policy, environmental guidelines such as those produced by the Forestry Commission, and standards. The designer may evaluate these or refine general objectives into rather

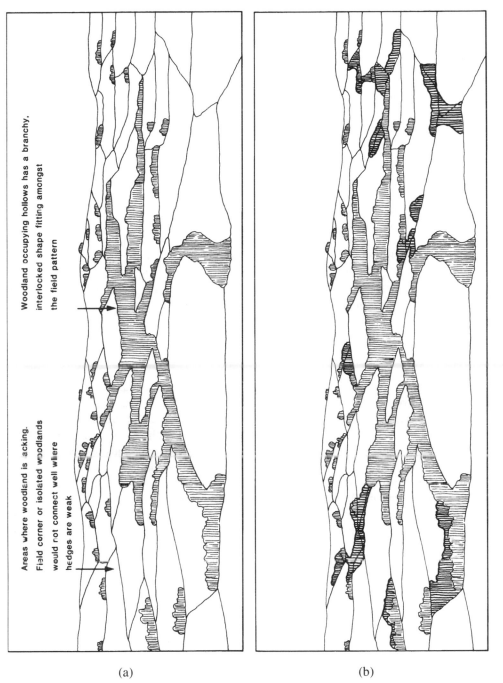

(a)   (b)

Figure 3.11. (a) Woodland in a rolling enclosed landscape. (b) Extra woodland expansion by 'growing' the pattern outwards. The interlocking pattern is maintained, connectivity in the landscape improved and its character respected. The pattern could grow up to a point of balance, where too much woodland begins to overwhelm the view or create too much enclosure for the scale of the landscape

more specific ones once he or she has analysed the landscape in question. If the landscape has a strong, intact, stable and healthy structure and pattern the designer is provided with something fairly easy to work with (Figure 3.11(a)). The pattern can be extended, infilled or otherwise 'grown' according to the respective influence of the elements which contribute to that pattern (Figure 3.11(b)). Photographs or sketches from the major viewpoints can be used as bases on which options are tried out. An analysis of the functional, ecological and aesthetic factors can be made to help determine the most appropriate options.

If the patterns are weaker, more creativity is possible, while in those situations where earlier patterns have completely disintegrated, whole new landscapes might be created. The inspiration to the designer may come from restoration ecology (Moffat and Buckley, Chapter 6), potential patterns from NVC classifications (Rodwell and Patterson, Chapter 5) or by borrowing metaphors from other areas of woodland ecology.

As well as the spatial layout, there is the time factor to be taken into account. We have already noted the dynamic nature of the landscape. New woodlands grow and other woodlands may be added which affect the overall composition. Modelling possible developments over several decades or even a century might help to determine how far it is desirable to go. The effect on the visual quality and the ecological functioning can be tested in this way, as can flows of timber and other outputs.

CONCLUSION

Woodland design in the landscape rarely starts with a blank canvas. The landscape has continuously evolved or changed and our actions are no more than the latest series of such changes. However, more recently, change has been faster and the resulting patterns less strongly rooted in the land, soil and climatic factors which constrained the possibilities in the past. More direction is needed if the right amounts and kinds of woodlands are to be created. A strategic approach based on a deep understanding of the landscape and its underlying dynamics will help determine the most appropriate locations, amounts and general design criteria. A more careful local analysis of proposed sites, backed up by knowledge of some basic aesthetic, ecological and practical skills, will convert the general strategy into suitable actual designs which achieve a goodness-of-fit with the landscape.

REFERENCES

Antikainen, M. (1993). *Forest Landscape Planning in Koli National Park*. Finnish Research Institute Research Paper 456, Helsinki.

Bell, S. (1993). *Elements of Visual Design in the Landscape*. E & F Spon, London.

Blunden, J. and Curry, N. (1988). *A Future for Our Countryside*. Blackwell, Oxford.

Brogan, H. (1985). *Longman History of the United States of America*. Longman, London.

Clarke, D.V., Cowie, T.G. and Foxon, A. (1985). *Symbols of Power at the Time of Stonehenge*. HMSO, Edinburgh.

Countryside Commission (1993). *Landscape Assessment: New Guidance*. CPP 423, Countryside Commission, Cheltenham.

Crowe, S. and Mitchell, M. (1988). *The Pattern of Landscape*. Packard Publishing, Chichester.

Department of the Environment (1992). *Indicative Forestry Strategies*. Circular 29/92. HMSO, London.

Fleming, A. (1988). *The Dartmoor Reaves*. Batsford, London.

Forestry Commission (1989). *Forest Landscape Design Guidelines*. HMSO, London.

Forestry Commission (1991). *Community Woodland Design Guidelines*. HMSO, London.

Forestry Commission (1992). *Lowland Landscape Design Guidelines*. HMSO, London.

Harrison, R.P. (1992). *Forests: the Shadow of Civilisation*. University of Chicago Press, Chicago.

Hill, M.O. (1979). TWINSPAN — a FORTRAN Program for Arranging Multivariate Data in an Ordered Two-way Table by Classification of the Individuals and Attributes. Section of Ecology and Systematics, Cornell University, Ithaca, New York.

Hoskins, W.G. (1955). *The Making of the English Landscape*. Hodder and Stoughton, London.

Hussey, C. (1967). *English Gardens and Landscapes*, 1700–1750. Country Life, London.

Lucas, O.W.P. (1991). *The Design of Forest Landscapes*. Oxford University Press, Oxford.

Norberg-Schulz, C. (1980). *Genius Loci*. Academy Editions, London.

Price, G. (1993). *Landscape Assessment for Indicative Forestry Strategies*. Forestry Authority, Cambridge.

Pyne, S.J. (1982). *Fire in America*. Princeton University Press, Princeton.

Rackham, O. (1986). *The History of the Countryside*. Dent, London.

Rodwell, J. and Patterson, G. (1994). *Creating New Native Woodlands*. Forestry Commission Bulletin 112. HMSO, London.

Steane, J.M. (1985). *The Archaeology of Medieval England and Wales*. Croom Helm, Beckenham.

Wilson, A. (1992). *The Culture of Nature: North American Landscape from Disney to the Exxon Valdez*. Blackwell, Oxford.

4 Biogeography and Woodland Design

I. F. SPELLERBERG

INTRODUCTION

Biogeography is the scientific study of the distribution of plants and animals in time and in space. It is an interdisciplinary science which draws on information from many sources (e.g. biology, ecology and geography) to describe and interpret patterns of plant and animal distribution, both past and present (Pears, 1977). In the context of woodland creation, biogeography might, for example, be used to explore some of the following:

- The shape of the woodland
- Area of woodland
- Position in the landscape
- Overall structure of the woodland (is it composed of smaller woodlands or groups of trees?)
- Features other than trees that are in the woodland.

But, as is the case with other disciplines, such opportunities to comment on biogeographical aspects should have clearly defined objectives. The establishment of a multi-use woodland for timber production, recreation and conservation of biological diversity could be an objective and biogeography could help to identify the spatial design of a multi-use woodland at several different scales.

This chapter starts by putting the case for conservation of biological diversity (biodiversity) into perspective with a particular emphasis on woodlands. The main content is an appraisal of the application of biogeography in the design of woodlands for the conservation of biodiversity. A brief outline of the misapplication of island biogeographical theory is thought to be necessary because the misapplication of island biogeographical theories continues to have a poor influence in conservation management strategies. A description is then given of some of the more recent aspects of biogeographical research of relevance to woodland design. Concerns about habitat fragmentation have led to particular regard being given to 'buffer zones' and 'wildlife corridors' as remedial treatments. These have been selected for specific attention because of their overall relevance to woodland design.

The Ecology of Woodland Creation. Edited by Richard Ferris-Kaan.

LOSSES IN BIODIVERSITY: CAUSES AND REMEDIES

The term 'biodiversity' (popularised by E.O. Wilson, 1985) emphasises that conservation of variety or diversity at different levels (genetic, population, species, habitats, ecosystems) is more important than just conservation of species. That there are declines in biodiversity throughout the world there is no debate. On land, pollution, exploitation and persecution have contributed to biodiversity loss, but by far the greatest cause has been habitat damage, habitat reduction and the fragmentation of habitats (Spellerberg, 1992). The concept of the process of habitat fragmentation is shown in Figure 4.1.

Habitat degradation

In some regions of the world, such as Europe and North America, this process of damage, loss and fragmentation of habitats has been going on for centuries. In Britain, habitat loss has taken place over hundreds of years and continues. For example, in the last 50 years over 80% of lowland meadow and chalk grassland has been lost and 30-50% (varies from county to county) of ancient semi-natural broadleaved woodland has been lost (RSNC, 1989).

In other regions such as in Madagascar, Indonesia and the Amazon basin, the process is more recent. There are now many pictorial examples showing the scale in both space and in time of habitat loss and fragmentation. The popular conception of the habitat fragmentation process is similar to a jigsaw puzzle and progressive loss of parts of the puzzle. In reality, the effects on biodiversity come from both deterioration in habitat quality and in complete loss of habitats. For example, interior woodland birds may be affected both by actual reduction and loss of woodland habitat and by deterioration of the woodland habitat. The result may be that some species are represented by much reduced populations in woodlands which in superficial appearance seem unchanged. In other words the habitat does not have to disappear (parts of the jigsaw need not be lost) to cause reduced and isolated populations.

The damage to habitats, reduction in area of suitable habitats and fragmentation of habitats may result in a patchy distribution of the remaining habitats. In other words the structural complexity or the heterogeneity of the environment may be increased. For some species, a patchy environment is not a disadvantage and may sometimes be beneficial. Some birds of prey, for example, find all the resources they require in a diverse landscape or a patchy environment because the patchiness or heterogeneity helps to support suitable levels of prey species over a large geographical scale. However, for many taxa which are not as mobile as birds of prey, the process of habitat degradation can lead, among other things, to reduced population levels and isolated populations. This may be particularly true of woodland invertebrates (see Key, Chapter 10).

In general terms the process of habitat damage, reduction and fragmentation could include all or any of the following:

Abiotic changes

● Change and/or reduction in area
● Perimeter of habitat fragment exposed to perturbations (pollution, physical damage)
● Distance to similar types of habitats increases
● Newly created edge causes changes in light, humidity and temperature levels

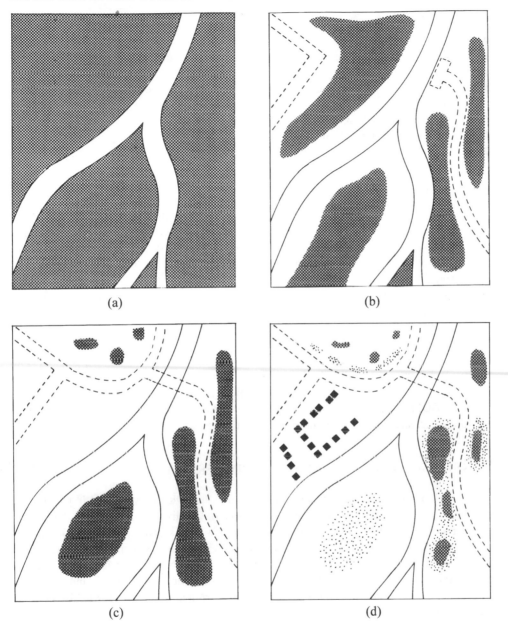

Figure 4.1. A conceptual series to show the process of woodland habitat loss and fragmentation over time: In (a) there is undisturbed woodland, with a river running through it. Over time there are changes in land-use and the woodland is reduced in area and fragmented. In (d) patches of original and natural woodland remain, some as small integral patches and some as patches within badly disturbed larger tracts of woodland

Biotic changes

- Newly created edge favours colonisation by new species
- Invasive and exotic species may penetrate the habitat fragment
- Change in the original number of species
- Extinction of keystone species leads to changes in species composition and extinction of other species
- Reduction in population levels
- Isolation of some populations
- Reduced genetic diversity within isolated populations.

Ecological processes

The ecological processes which are initiated by habitat fragmentation have not been fully studied, although more recently there has been a lot of interest in this area of ecology. Consequently there are some observations, particularly from studies on birds, which seem to support some of the above. For example, Diamond (1973) has described the effects of forest fragmentation on birds in New Guinea, where several species became extinct. Opdam *et al.* (1984), working on woodland birds in the Netherlands, have shown that, apart from areas of woodland fragment, the amount of woodland nearby and distance to large woodland areas have significant effects on woodland bird species richness (see also Fuller *et al.*, Chapter 11). Soule *et al.* (1988) have shown that chaparral birds may be even more prone to extinction in fragmented landscapes because of their poor dispersal ability and because the distribution of native predators may also influence species numbers. More recently, results from the 'Biological Dynamics of Forest Fragments Project' (BDFF) in the Amazon have shown what happens to biological communities as the forest is cleared, leaving behind a few forest remnants (Lovejoy *et al.*, 1986).

CONSERVATION OF WOODLAND BIODIVERSITY

Afforestation and the growth of forest monocultures have been responsible in part for the concern expressed previously about low levels of biodiversity in forest plantations. Consequently there has been much research on how levels could be increased and it is now clear that there are many ways by which levels of biodiversity might be increased in woodlands. There have been very important initiatives in North America, Britain and in Scandinavia, where ways of increasing levels of biodiversity in woodlands and plantation forests have been examined (e.g. Thomas, 1979; Marion and Werner, 1987; Mitchell and Kirby, 1989; Hunter, 1990; Bunnell *et al.*, 1992).

Assessment of biodiversity in Britain

In 1990–91, the UK Forestry Authority commissioned a programme of research on various aspects of multi-use forests including the landscape aspects, the valuation of non-timber products and conservation of biodiversity. As part of that programme, research was undertaken at the Centre for Environmental Sciences, University of Southampton, to review the methods previously and currently used for increasing levels of biodiversity in forest plantations. On the basis of an extensive literature review (Spellerberg and Sawyer,

1993) it appeared that the following aspects of forest ecology could potentially be managed to increase levels of biodiversity:

- Extent of open space
- Extent of dead and dying wood
- Tree species richness
- Age class of trees and retention of very old trees
- Silviculture
- Ground vegetation
- Ride edge structure and vegetation
- Riparian zones and aquatic habitats
- Artificial resources such as nest boxes.

Setting and monitoring standards

As is the case in all living communities, a woodland community will change as a result of ecological succession and therefore the quality and distribution of wildlife habitats will change over time. There will therefore be opportunities for habitat management. The creation of habitats and management of habitats will be undertaken with certain objectives in mind; these objectives could be expressed in various ways such as levels of species richness or increases in population density for certain species. It is likely that such objectives may in the future be coupled with standards.

The UK Forestry Authority has taken a lead in establishing standards for forest management (Forestry Commission, 1989; 1990; 1991) and more recently has looked at ways of defining and quantifying standards for levels of biodiversity in forests. This trend towards establishing standards for biodiversity seems likely to be adopted by forest authorities in Europe, Scandinavia and North America.

It would seem wrong to design and undertake habitat creation and habitat management without appraising the degree of success (or failure) of the work. It would also seem wrong to adopt standards for biodiversity without undertaking some kind of assessment to see if the standards are being achieved. That being the case, it would seem necessary to establish an ecological monitoring programme, designed to accumulate appropriate data over certain periods of time. There are now well-established cost-effective methods for ecological monitoring programmes (Spellerberg, 1991a).

ISLAND BIOGEOGRAPHICAL THEORIES

The area of woodland, the extent of structural complexity of the woodland, the shape of the woodland, the location of the woodland in the landscape and the nature of the surrounds could all have implications for the ecology of the wildlife and levels of biodiversity in the woodland. For decades, the relevance of woodland shape, size and location to wildlife conservation has been the subject of extensive research, much of which was poorly conceived by drawing comparisons between real islands and so-called 'habitat islands'.

Species – area relationships

Research on island biogeography has produced many theories. Since at least 1962, that research has focused on the simple observation that for certain taxonomic groups there is generally an increase in species richness with an increase in island area. Williams (1964) suggested that the number of species on an island was determined by the variety of habitats and because larger islands tend to have a greater variety of habitats, more species will be supported on larger islands than small islands. Species richness is the number of species whereas species diversity is, strictly speaking, the relative abundance of species (Spellerberg, 1991a). There are many studies which have shown increasing species richness with increasing island area, but in many cases it has been concluded that area is probably only one of a number of possible variables (see, for example, the following: Toft and Schoener, 1983; Gardner, 1986; Harris, 1973; Quinn *et al.*, 1987).

The relationship between area of the island and the number of species can be expressed in the form of

$$S = CA^k$$

where S is the number of species
C is a constant
A is the area
k is the slope

On a plot of the log value of species against the log value of the area, the value of k for real islands has been found to be most commonly between 0.2 and 0.3 (when $k = 0.3$, an island ten times as large as a small island would have about twice as many species). For habitat fragments, k is often slightly less in value.

Modelling island population dynamics

A milestone in island biogeographical theories came in 1963 when MacArthur and Wilson published their classic paper, then later in 1967 their book, *The Theory of Island Biogeography* (MacArthur and Wilson, 1963; 1967). They looked at many aspects of island biogeography, one of which was what determined the number of species on an island. A dynamic equilibrium model was suggested in which the rates of immigration and rates of extinction on an island determined the equilibrium number of species.

In the years that followed, many people have tried to find evidence in support of the dynamic equilibrium model and other theories by way of field observation and laboratory experiments. There is now much evidence (from both field and laboratory studies) to show that for certain taxonomic groups there is a logarithmic relationship between island area and the number of species but little evidence to support the Equilibrium Theory as a whole (Spellerberg, 1991b). In general, it would seem that species richness of some taxonomic groups on some islands is determined in part by several variables including the following:

- Topographical and structural diversity of the island
- The variety of habitats
- Area
- Distance from source of colonization
- Species richness at the source
- Area of land providing the source.

At the same time there was much interest in using these theories of island biogeography in attempts to try and understand the ecology of fragmented habitats. There has been little success in such attempts and there has been much misinterpretation or bad reporting of what exactly the various theories of island biogeography were about. A useful rule is to go back to the original account rather than accepting second-hand accounts. Looking back through the literature (there is a huge amount on island biogeography and conservation) it seems strange that any such attempt should have been made, considering that the ecology of an island surrounded by sea is hardly likely to provide a basis for comparison with the ecology of a habitat fragment which is an integral part of the landscape.

HABITAT FRAGMENT BIOGEOGRAPHY

Metapopulation processes

Not all populations of plants and animals consist of one large population. Many consist of subpopulations distributed throughout a variable or patchy environment. A population which is made up of subpopulations, interconnected via patterns of gene flow, extinction and recolonisation, is known as a metapopulation. The use of the term metapopulation commenced in the late 1960s and in 1970 when it was introduced by Levins and was first used to describe the ecology of populations (Levins, 1969; 1970). With so many habitats becoming fragmented, it is perhaps not surprising that the term metapopulation has become popular in conservation biology and in biogeography; many species which may formerly have had a fairly continuous distribution are becoming metapopulations because of habitat fragmentation. It has become important therefore to consider the area of the habitats, the effects of perturbations from the surrounding area, the spatial relationships with similar habitat patches and other kinds of habitats, and the conditions between habitats which may facilitate or reduce dispersal.

Habitat island size

Inspired by species–area relationships found on islands, much of the research on biogeography of fragmented habitats has focused on the number of species (of a particular taxonomic group) in different sized fragments. Misapplications of this research led to the popular rule that nature reserves should be as large as possible (the larger the area the more species). However, it was not long before this rule was challenged by some researchers who were able to give examples in which collections of small habitat fragments supported more species (of a particular taxonomic group) compared to a single habitat fragment of the same total area (Spellerberg, 1991b). Beyond this 'science' there was also the matter of management and there are arguments for and against single large reserves and several small reserves. This acrimonious discussion culminated in the SLOSS debate: Single Large or Several Small. What seems to have escaped the attention of this debate was the fact that there was variation; that is, variation in the relationship between species richness and area. In some instances small habitat fragments could support more species than might be expected, and conversely a large habitat could be found to support fewer species than might be predicted from a species–area curve for a particular taxonomic group. It is a generalisation to say that for a particular taxonomic group, the number of species tends to increase with increasing area of habitat. To base conservation management strategies on such a generalisation would seem to be unwise because of variation and exceptions. From a management perspective, increasing the number of species (all species, or species of a particular taxonomic group) may not always be appropriate. Conservation strategies based entirely on the idea of more species the better, without due consideration to other ecological considerations, would therefore not seem to be the best way to proceed; put simply, be wary of generalisations.

Habitat island diversity

Many hundreds of researchers have reported species–area curves for various taxonomic groups; for example, Peterken (1974: woodland flora); Shreeve and Mason (1980: woodland butterflies). Most studies of species richness of habitat fragments come from studies on birds. Examples for woodland and forest birds include Diamond (1973), Yapp (1979), Blake and Karr (1984), Robbins *et al.* (1989), Harms and Opdam (1990). Some of these investigations in Britain have gone beyond reporting the simple species–area relationship and have found that in some circumstances habitat diversity rather than area is better correlated with species richness (Rafe *et al.,* 1985). Of considerable conservation importance has been the observation that small woodland fragments support many edge dwelling bird species and that forest interior species are poorly represented in small forests (Blake and Karr, 1984). This has important implications in the UK where there are few 'true' forest interior species. In North America, Robbins *et al.* (1989) found that in relatively undisturbed mature forests the degree of isolation and area were significant predictors of relative abundance for more bird species than were any habitat variables.

SOLUTIONS TO HABITAT FRAGMENTATION

Two possible remedies for minimising the deleterious effects of habitat fragmentation are the development and/or maintenance of ecological buffer zones and wildlife corridors (Figure 4.2).

Ecological buffer zones

Buffer zones have been defined as regions adjacent to protected areas, on which land-use is restricted and which gives additional protection to the protected area itself and helps to maintain the ecological integrity of the interior of the protected area (Spellerberg, 1991b; Angold, 1992). The concept of buffer zones in conservation is not new and has notably been developed in connection with the management of national parks which have been established for large mammals (MacKinnon *et al.*, 1986). In temperate regions, the concept of ecological buffer zones has been implemented as long ago as 1974 when, in Michigan, the effectiveness of such zones was considered to be dependent upon local conditions, including the vegetation type in the buffer zone, the nature of the boundary (river, fence, etc.), the compatibility of the adjacent land use and expected degree of encroachment (Tans, 1974).

The buffer zone concept has also been used in woodland management. For example, in North America, Harris (1984) has drawn attention to the potential role of mature stands

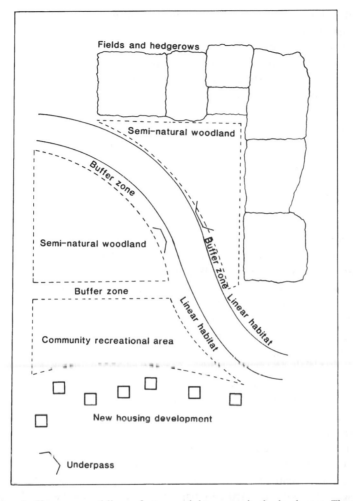

Figure 4.2. How buffer zones and linear features might appear in the landscape. There is a buffer zone between the community recreation area in the woodland. There is also a buffer zone (and linear and habitat) between the road and the woodland. Linear features exist in the form of hedgerows (some fragmented, some continuous) and in the form of an underpass under the road

of timber (around old growth ecosystems) for buffering against climate, fire and other perturbations. In Britain, the concept was used to identify a Heritage Area around the New Forest in the South of England. This Heritage Area consists of land similar to the major New Forest habitats and also includes land which is very important to the traditional management of semi-natural New Forest habitats. It was intended that this area would buffer the New Forest from change by:

1. Restricting industrial development on the forest edge.
2. Providing lands for stock which may not be left out in the forest over the winter.

The concept of buffer zones could usefully apply to multi-use woodlands. For example, on a large scale it may be desirable to minimise the effects of different surrounding land-use (agricultural, roads, houses or recreational areas) on the woodland. In those circumstances, it may seem desirable therefore to identify an area on the woodland perimeter and manage it accordingly with the object of buffering the woodland from the likely disturbance or pollution from the various surrounds. On a smaller scale, populations of some woodland species (such as woodland interior bird species) may become reduced and isolated as a consequence of habitat deterioration (not necessarily habitat loss) which has been caused by physical disturbance in the woodland. It may be desirable, therefore, to establish buffer zones around the core woodland areas. However, while the concept of a buffer zone seems sensible, there has so far been little in the way of research on the ecology of buffer zones.

Wildlife corridors

Linear landscape features such as hedgerows, avenues of trees, roadside verges, woodland rides, fire breaks, cleared areas along powerline routes, lake shores, canal edges and river banks can all provide linear habitats and could all act as conduits for movement of wildlife from one habitat to another. With increasing concern about the loss, fragmentation and isolation of habitats, attention has been focused on ways of reducing isolation effects. The concept that linear landscape features (Figure 4.2) could be corridors for the dispersal of wildlife is not new and has been introduced in connection with many conservation projects: bamboo forest corridors for giant pandas in China; forest corridors for elephants in Sri Lanka; rainforest corridors for golden-headed tamarins in Brazil (Spellerberg and Gaywood, 1993). The corridor concept has now become so popular that it is often assumed that any linear feature will automatically be a corridor. Local authorities have commonly incorporated the idea into structure plans and have adopted the name of greenways to indicate that linear features can have an amenity value as well as a wildlife corridor function.

Important and interesting as the concept may be, there has been a tendency to overlook the need for supporting evidence: does wildlife actually move from one population to another along linear features? Is there evidence to support the view that certain kinds of linear landscape features can facilitate dispersal of wildlife and help to lessen the effects of isolation?

In a report for English Nature (Spellerberg and Gaywood, 1993), the literature on linear landscape features was reviewed and it was concluded that linear features can be grouped broadly into three categories:

1. *Biological:* habitats, corridors
2. *Environmental:* wind breaks, etc.
3. *Amenity.*

It was evident that there had been much research to show that linear landscape features such as woodland rides can provide important linear or edge habitats. For some animal groups, there was evidence to show that dispersal could take place along linear landscape features. However, as yet, there has been very little research which has been designed to look at the role of linear landscape features in either contributing to valuable landscape features or reducing population isolation effects brought about by fragmentation of habitats.

The role of ride edges in woodlands as linear wildlife habitats and corridors has been investigated for plants and several animal taxonomic groups including butterflies

(Warren, 1985; Ferris-Kaan and Patterson, 1992; Greatorex-Davies *et al.*, 1993), reptiles (Dent and Spellerberg, 1988) and small mammals (Gurnell, 1985). The potential role of rides as habitats and conduits for dispersal within a woodland environment must surely be taken into consideration in the design of woodlands, and fortunately the results of some research can provide very specific data on which to design ride edges. For example, the width, shape, orientation and vegetation composition are just some of the parameters which have been put into practice (Carter and Anderson, 1987; Warren and Fuller, 1993).

The growth of the trees throughout the woodland combined with varied uses within the woodland will result in changes in the quality of wildlife habitats. Human intervention would, therefore, seem to be justified because it would be important to try and model the possible changes which may occur to ride habitats and plan management accordingly. Modelling provides an important basis for predicting the length of time for which the habitats will be suitable for certain species. For example, work on the sand lizard (*Lacerta agilis*) (a protected species) on forest ride edges in plantations in the south of England included research on modelling the changes in the quality of the habitat (Dent and Spellerberg, 1988). This model has important applications for conservation of the lizard species concerned but, more importantly, the same model could be used for other small vertebrates occupying ride edge habitats in forest plantations and in woodlands.

SUMMARY

Aspects of biogeography could have useful applications when creating multi-use woodlands, especially if one of the functions of the woodland is conservation of biodiversity. For example, a biogeographical approach would be appropriate when considering the position of the woodland in the landscape and its relation to other woodlands, the spatial configuration of the woodland and the internal structure of the woodland (and variation over time).

The large amount of research on island biogeography and its application to conservation of mainland habitats includes some misinterpretation of island biogeographical theory. Perhaps of greater relevance is the view that it would seem illogical to even attempt to draw on comparisons between island biogeography and the biogeography of wildlife in a mosaic of mainland habitats. It is concluded therefore that it is important to clearly differentiate between island biogeography and the biogeography of habitat fragments.

The wildlife habitats in a newly created woodland would presumably be designed and managed on the basis of the theories and empirical studies derived from research on the biogeography of wildlife in a landscape made up of a mosaic of habitats. Metapopulation theory draws attention to subsets of populations living in a patchy environment where there is emigration and extinctions among the subpopulations. Some taxa will depend on a patchy environment, but there has been concern over the rate at which habitats have become fragmented with resulting isolation of some populations. A new woodland should, therefore, be designed with some consideration given to means of dispersal (of the regional fauna and flora) between the woodland and other woodlands in the area. Similarly, the internal structure of the woodland could usefully be designed with some thought being given to spatial configuration and size.

There has been much fascination with the apparent link between area and number of species to the extent that cause and effect has been the subject of less rigorous research.

In general, and for some taxa in woodlands, species–area curves can be found, but there may be other variables which need to be considered. It would seem wise to avoid generalisation such as the larger the better on the basis of simple observations of species–area curves.

Ecological buffer zones and landscape linear features are relevant to the spatial and temporal patterns of wildlife distribution and movement. Woodland creation provides opportunities to research both of these topics, which previously have become widely accepted with less than desirable empirical research. Both topics could usefully be researched at many scales (from the whole landscape to woodland rides).

With the creation of new woodlands, there would seem to be very good opportunities to establish long-term monitoring programmes which in themselves could be productive in terms of the biological data being recorded. Ecological monitoring programmes would also provide a very important basis for appraising the success of habitat creation and success of habitat management.

Acknowledgements
I am grateful to John Sawyer for his help in researching the literature and to Richard Ferris-Kaan and Jenny Claridge for their help in editing and improving the text.

REFERENCES

Angold, P.G. (1992). The role of buffer zones in the conservation of semi-natural habitats. PhD Thesis, University of Southampton, 213 pp.

Blake, J.G. and Karr, J.R. (1984). Species composition of bird communities and the conservation benefit of large versus small forests. *Biological Conservation*, **30**, 173–187.

Carter, C.I. and Anderson, M.A. (1987). *Enhancement of Lowland Forest Ridesides and Roadsides to Benefit Wild Plants and Butterflies*. Forestry Commission Research Information Note 126. Forestry Commission, Farnham.

Bunnell, F.L., Daust, D.K., Klenner, W., Kremsater, L.L. and McCann, R.K. (1992). Managing for Biodiversity in Forested Ecosystems. Report to the Forest Sector of the Old Growth Strategy. Land Management Report. British Columbia Ministry of Forests.

Dent, S. and Spellerberg, I.F. (1988). Use of forest ride verges in southern England for the conservation of the sand lizard *Lacerta agilis* L. *Biological Conservation*, **45**, 267–277.

Diamond, J.M. (1973). Distributional ecology of New Guinea birds. *Science*, **179**, 759–769.

Ferris-Kaan, R. and Patterson, G.S. (1992). *Monitoring Vegetation Changes in Conservation Management of Forests*. Forestry Commission Bulletin 108. HMSO, London.

Forestry Commission (1989). *Forest Landscape Design Guidelines*. HMSO, London.

Forestry Commission (1990). *Forest Nature Conservation Guidelines*. HMSO, London.

Forestry Commission (1991). *Forests and Water Guidelines*. HMSO, London.

Gardner, A.S. (1986). The biogeography of lizards of the Seychelles Islands. *Journal of Biogeography*, **13**, 237–253.

Greatorex-Davies, J.N., Sparks, T.H., Hall, M.L. and Marrs, R.H. (1993). The influence of shade on butterflies in rides of coniferised lowland woods in southern England and implications for conservation management. *Biological Conservation*, **63**, 31–41.

Gurnell, J. (1985). Woodland rodent communities. *Symposium of the Zoological Society of London*, **5**, 377–411.

Harms, W.B. and Opdam, P. (1990). Woods as habitat patches for birds: application in landscape planning in the Netherlands. In: Zonneveld, O. and Forman, R.T.T. (eds.). *Changing Landscapes: an Ecological Perspective*. Springer, New York, 73–97.

Harris, L.D. (1984). *The Fragmented Forest. Island Biogeographical Theory and the Preservation of Biotic Diversity*. University of Chicago Press, Chicago.

Harris, M.P. (1973). The Galapagos avifauna. *Condor*, **75**, 265–278.

Hunter, M. (1990). *Wildlife, Forests and Forestry: Principles of Managing for Biological Diversity*. Prentice Hall, Englewood Cliffs, New Jersey.

Levins, R. (1969). Some demographic and genetic consequences of environmental heterogeneity for biological control. *Bulletin of the Entomological Society of America*, **15**, 237–240.

Levins, R. (1970). Extinction. In: Gerstenhaber, M. (ed.). *Some Mathematical Problems in Biology*. R.I. American Mathematical Society, Providence, 77–107.

Lovejoy, T.E., Bierregaard, R.O., Rylands, A.B., Malcolm, J.R., Quintela, C.E., Harper, L.H., Brown, K.S., Powell, A.H., Powell, G.V.N., Schubart, H.O.R. and Hays, M.B. (1986). Edge and other effects of isolation on Amazon forest fragments. In: Soule, M.E. (ed.). *Conservation Biology: the Science of Scarcity and Diversity*. Sinauer, Massachusetts, 257–285.

MacArthur, R.H. and Wilson, E.O. (1963). An equilibrium theory of insular zoogeography. *Evolution*, **17**, 373–387.

MacArthur, R.H. and Wilson, E.O. (1967). *The Theory of Island Biogeography*. Princeton University Press.

MacKinnon, J., Mackinnon, K., Child, G. and Thortsell, J. (1986). Managing protected areas in the tropics. IUCN, Gland, Switzerland.

Marion, W.R. and Werner, M. (1987). Management of Pine Forests for Selected Wildlife in Florida. Florida Cooperative Circular 706, University of Florida, Gainesville, Florida.

Mitchell, P.L. and Kirby, K.J. (1989). *Ecological Effects of Forestry Practices in Long Established Woodland and Their Implications for Nature Conservation*. Oxford Forestry Institute Occasional Paper No. 39, University of Oxford, Oxford.

Opdam, P., Dorp, van D. and Braakter, C.J.F. (1984). The effect of isolation on the number of woodland birds in small woods in the Netherlands. *Journal of Biogeography*, **11**, 473–478.

Pears, N. (1977). *Basic Biogeography*. Longman, London.

Peterken, G.F. (1974). A method for assessing woodland flora for conservation using indicator species. *Biological Conservation*, **6**, 239–245.

Quinn, S.L., Wilson, J.B. and Mark, A.F. (1987). The island biogeography of Lake Manpouri, New Zealand. *Journal of Biogeography*, **14**, 569–581.

Rafe, R.W., Usher, M.B. and Jefferson, R.G. (1985). Birds on reserves: the influence of area and habitat on species richness. *Journal of Applied Ecology*, **2**, 327–335.

Robbins, C.S., Dawson, D.K. and Dowell, B.A. (1989). Habitat area requirements of the breeding forest birds of the middle Atlantic states. *Wildlife Monographs*, **103**, 1–34.

Shreeve, T.G. and Mason, C.F. (1980). The number of butterfly species in woodlands. *Oecologia*, **45**, 414–418.

RSNC (1989). *Losing Ground*. Royal Society for Nature Conservation, Lincoln.

Soule, M.E., Bolger, D.T., Alberts, A.C., Wright, J., Sorice, M. and Hill, S. (1988). Reconstructed dynamics of rapid extinctions of chaparral-requiring birds in urban habitat islands. *Conservation Biology*, **2**, 75–92.

Spellerberg, I.F. (1991a). *Monitoring Ecological Change*. Cambridge University Press, Cambridge.

Spellerberg, I.F., (1991b). Biogeographical basis of conservation. In: Spellerberg, I.F., Goldsmith F.B. and Morris, M.G. (eds). *The Scientific Management of Temperate Communities for Conservation*. Blackwell Scientific Publications, Oxford, 293–322.

Spellerberg, I.F. (1992). *Evaluation and Assessment for Conservation*. Chapman and Hall, London.

Spellerberg, I.F. and Gaywood, M. (1993). *Linear Features: Linear Habitats and Wildlife Corridors*. English Nature Research Reports No. 60. English Nature, Peterborough.

Spellerberg, I.F. and Sawyer, J.W.D. (1993). Biodiversity in Plantation Forests: Increasing Levels and Maintaining Standards. Report to the Forestry Authority, UK.

Tans, W. (1974). The priority ranking of biotic natural areas. *The Michigan Botanist*, **13**, 31–39.

Thomas, J.W. (1979). *Wildlife Habitats in Managed Forests, the Blue Mountains of Oregon and Washington*. Technical edition: Agriculture Handbook 553. USDA Forest Service, Washington, DC.

Toft, C.A. and Schoener, T.W. (1983). Abundance and diversity of orb spiders on 106 Bahamian Islands: biogeography at an intermediate tropic level. *Oikos*, **41**, 411–426.

Warren, M.S. (1985). The influence of shade on butterfly numbers in woodland rides with special reference to the woodland white *Leptidae sinapis*. *Biological Conservation*, **33**, 147–164.

Warren, M.S. and Fuller, R.J. (1993). *Woodland Rides and Glades: Their Management for Wildlife*. English Nature, Peterborough.

Williams, C.B. (1964). *Patterns in the Balance of Nature and Related Problems of Quantitative Ecology*. Academic Press, New York.
Wilson, E.O. (1985). The biological diversity crisis. *Bioscience*, **35**, 700–705.
Yapp, W.B. (1979). Specific diversity in woodland birds. *Field Studies*, **5**, 45–58.

5 Vegetation Classification Systems as an Aid to Woodland Creation

J. S. RODWELL and G. S. PATTERSON

INTRODUCTION

Understanding the patterns of variation among existing kinds of broadleaved woodland is essential for the sensitive creation of new woods intended to have a more natural character. Classifications of vegetation can help by furnishing descriptions of different kinds of woodland already present as familiar and appreciated elements of our landscape and able to serve as models for planting and management programmes. Some classifications of British woodlands have concentrated on older stands, characterising a range of ancient woodland types (Rackham, 1980; Peterken, 1981) and giving a picture of 'past-natural' variation upon which we might wish to draw. Other schemes, like the National Vegetation Classification (NVC), have had a broader remit, sampling woods of all ages and origins together and describing woodland communities which represent a fuller realm of 'present-natural' variation (Rodwell, 1991a). In these sorts of approaches, which purport to be natural classifications emerging from the assembled data themselves, rather than being artificially imposed taxonomies of vegetation types, the mixtures of species described are recognised as plant communities with a measure of integrity and stability, parts of complex ecosystems which persist in a dynamic but sustained fashion in balance with their landscape context. Clearly, attempting to create such systems entirely from scratch is a tall order (see Spencer, Chapter 1). Long surviving stands of woodland, with their great complexity of composition and structure, their intricate networks of relationships with the physical habitat and associated fauna, and their accumulation of historical, socio-economic and cultural associations, cannot be reproduced. Even secondary woods, arising by natural regeneration, may often need many decades before they acquire a reasonably full quota of their characteristic plants (Peterken and Game, 1984).

However, native trees are being increasingly planted in Britain for a wide range of objectives, and there are frequently opportunities for thinking in terms of encouraging mixtures of native trees and shrubs appropriate to particular sites in the hope of approaching the appearance and ecological integrity of existing woodland canopies and understoreys. Such mixtures can be selected using knowledge of the ecological preferences of individual trees and shrubs (Soutar and Peterken, 1989a,b), but classifications are

The Ecology of Woodland Creation. Edited by Richard Ferris-Kaan.
© 1995 The editor, contributors and the Forestry Authority. Published in 1995 by John Wiley & Sons Ltd.

especially helpful in providing a general background understanding of species assemblages among existing woods and can help build confidence in the attainability of the goal of creating new stands. More particularly, classifications can help:

- Predict the kind of woodland vegetation which we might expect to develop on a site if succession were to proceed unhindered.
- Select lists of ecologically appropriate trees and shrubs for planting in such situations.
- Identify optimal precursor vegetation characteristic of the site, whose herbaceous species could give a head start to developing a woodland ground cover.
- Indicate desired invaders among the prospective ground cover, whose appearance could be monitored and encouraged.
- Develop an ecological basis for the design of pattern among woodland types over diverse landforms.
- Suggest styles of management for the establishment phase and beyond, which could increase the likelihood of success in creating the new woodlands.

This chapter illustrates how each of these processes might be informed, both by the use of classifications in general and with particular reference to the NVC, an approach recently developed by Rodwell and Patterson (1994).

PREDICTING APPROPRIATE PLANTING MIXTURES

The development of the National Vegetation Classification

If classifications of woodland vegetation are developed independently of environmental data, simply by the analysis of plant species records from sample locations, then subsequent correlation of groups of samples with habitat factors can be a powerful corroboration of the ecological meaning of the emerging woodland types. In the NVC, for example, floristic data were collected in a systematic fashion from over 2800 localities throughout Britain and analysed using a variety of multivariate computational techniques. A total of 19 major types of woodland vegetation was recognised, each community characterised by distinctive mixtures of trees and shrubs associated with characteristic herbs and, often too, some ferns, mosses, liverworts and lichens. Each of these communities was shown to be related to particular climatic and edaphic conditions and represents the kind of vegetation that could develop wherever such conditions were fulfilled, were succession allowed to progress towards its conclusion. Although such successions have often been modified by man and the resulting stands greatly influenced by management, interference and neglect, these woodlands still provide important clues as to the kinds of vegetation we might aim for in planting broadleaved woodlands where the intention is to encourage a natural character. More particularly, classifications of these woodlands enable us to predict appropriate mixtures of trees and shrubs suitable for existing climatic and soil conditions in sites which are at present unwooded.

Mixed broadleaved woodlands in Britain

For example, among the mixed broadleaved woodlands of Britain, there is an NVC type which could conveniently be described as lowland bluebell woodland (community W10 *Quercus-Pteridium-Rubus* woodland in Rodwell, 1991a). Oak, mostly *Quercus robur* with *Q. petraea* locally important, and birch, usually *Betula pendula* with *B. pubescens* more occasionally, are the usual dominant trees, with small-leaved lime (*Tilia cordata*), hornbeam (*Carpinus betulus*) and the introduced sweet chestnut (*Castanea sativa*) locally prominent in the south-east, ash (*Fraxinus excelsior*), wych elm (*Ulmus glabra*) and the introduced sycamore (*Acer pseudoplatanus*) to the north and west. Hazel (*Corylus avellana*) and hawthorns (*Crataegus monogyna* and the rare *C. laevigata*) are the commonest shrubs with holly (*Ilex aquifolium*), rowan (*Sorbus aucuparia*), crab apple (*Malus sylvestris*), elder (*Sambucus nigra*) and a variety of other species also represented occasionally in the understorey.

Such woodland vegetation is confined to the warmer and drier lowlands of Britain and occurs typically on brown earths of neutral or moderately acidic reaction and more base-poor ground-water gleys, derived from non-calcareous sedimentary shales and clays and lime-poor superficials like clay-with-flints, more ill-draining sands-and-gravels and old alluvium. Thus, wherever such habitat conditions coincide, as in the gently undulating basins, vales and low plateaux in central and eastern England, and along valley bottoms and sides around the fringes of the uplands to the north and west, the species assemblages encountered in surviving stands could provide the basis for selecting planting mixtures in new woodlands.

Planting zones

Using this kind of predictive approach (Rodwell and Patterson, 1994), it is possible to delimit five major planting zones in Britain, reflecting for the most part gross climatic differences across the country, such as contrasts in patterns of temperature and rainfall (Figure 5.1). Within such zones, it is then possible to categorise the terrain and soil types at a particular site and predict the trees and shrubs most appropriate for planting. With wetter woodlands, planting zones can be defined on the basis of mapping suitable ground conditions, whose character tends to override climate as the determinative control of distribution. In all, of the 19 woodland communities characterised by the NVC, Rodwell and Patterson (1994) provided predicted mixtures for 15 types: six types of oak-birch (*Quercus-Betula*) and mixed broadleaved woodland, three types of beech (*Fagus sylvatica*) woodland, pine (*Pinus sylvestris*) woodlands, juniper (*Juniperus communis*) woodlands and four kinds of wet woodland with alder (*Alnus glutinosa*), downy birch (*Betula pubescens*) and willows (*Salix* spp.).

SELECTING FROM AMONG PREDICTED SPECIES

Some of the trees and shrubs recommended for planting might, of course, be already present on a site as a result of natural colonisation and not all the species found in existing stands need be planted in every new woodland of the selected type; nor need the species be planted everywhere in the same proportions. Indeed, classifications can help categorise trees and shrubs as possible major or minor elements in planting programmes on the basis of their frequency and abundance in existing stands (Figure 5.2).

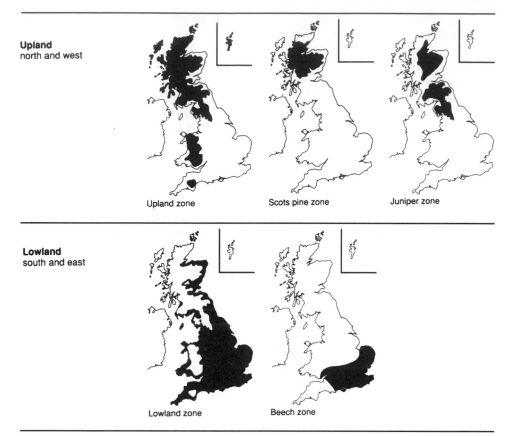

Figure 5.1. Major planting zones for new native woodlands in Britain. Reproduced from J. Rodwell and G. Patterson (1994), *Creating New Native Woodlands*, Forestry Commission Bulletin 112, by permission of HMSO

Planting patterns

Individual types of semi-natural woodland themselves show a wealth of variation in the proportions of their characteristic species and there can be ample scope in planting new stands for imaginative variation in design, both horizontally across the planted area and vertically among prospective canopy and understorey elements. Where timber growing was an objective in management, for example, the proportion of more productive trees could be increased.

Combinations of planted clumps and open areas could be used with variation of composition and spacing within clumps and distance between clumps to encourage diversity of cover (Figure 5.3). However, ambitious hopes informed by classifications of existing woodland types should also be tempered by silvicultural experience of the difficulties of getting tree and shrub mixtures established (see Harmer and Kerr, Chapter 8). This can often be quite difficult and relies upon carefully timed tending operations in order to maintain components. Generally, it may be prudent to plant single-species clumps varied in size and shape to develop canopy diversity in the long term. Monitoring the success of planting programmes based on approaches of this kind is essential so that methods can be refined.

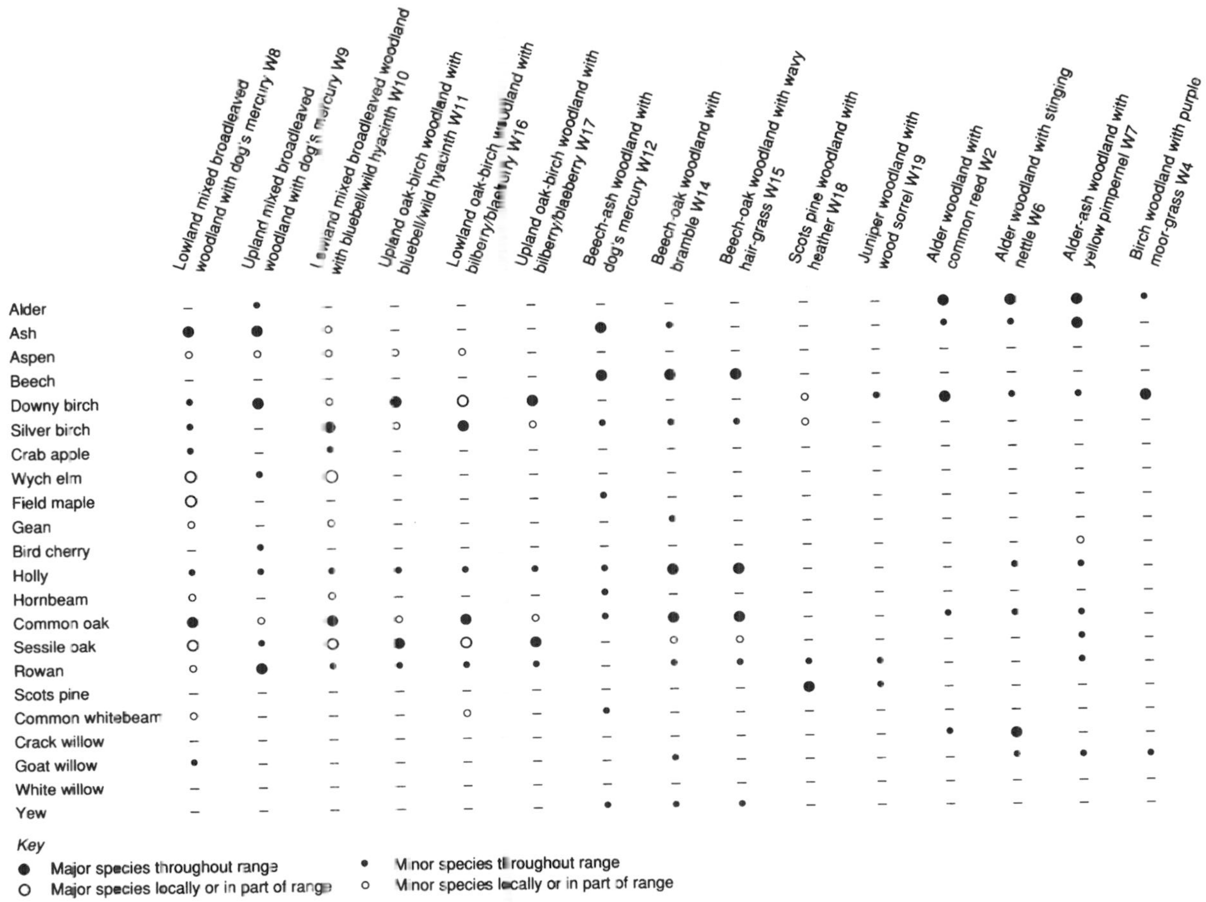

Part 1

Column headers (woodland types):
- Birch woodland with purple moor-grass W4
- Alder-ash woodland with yellow pimpernel W7
- Alder woodland with stinging nettle W6
- Alder woodland with common reed W2
- Juniper woodland with wood sorrel W19
- Scots pine woodland with heather W18
- Beech-oak woodland with wavy hair-grass W15
- Beech-oak woodland with bramble W14
- Beech-ash woodland with dog's mercury W12
- Upland oak-birch woodland with bilberry/bleaberry W17
- Lowland oak-birch woodland with bilberry/bleaberry W16
- Upland oak-birch woodland with bluebell/wild hyacinth W11
- Lowland mixed broadleaved woodland with bluebell/wild hyacinth W10
- Upland mixed broadleaved woodland with dog's mercury W9
- Lowland mixed broadleaved woodland with dog's mercury W8

Row labels (species):
Alder, Ash, Aspen, Beech, Downy birch, Silver birch, Crab apple, Wych elm, Field maple, Gean, Bird cherry, Holly, Hornbeam, Common oak, Sessile oak, Rowan, Scots pine, Common whitebeam, Crack willow, Goat willow, White willow, Yew

Key
● Major species throughout range
○ Major species locally or in part of range
• Minor species throughout range
o Minor species locally or in part of range

Figure 5.2. Major and minor species of trees and shrubs in existing semi-natural woodlands. Reproduced from J. Rodwell and G. Patterson (1994), *Creating New Native Woodlands*, Forestry Commission Bulletin 112, by permission of HMSO

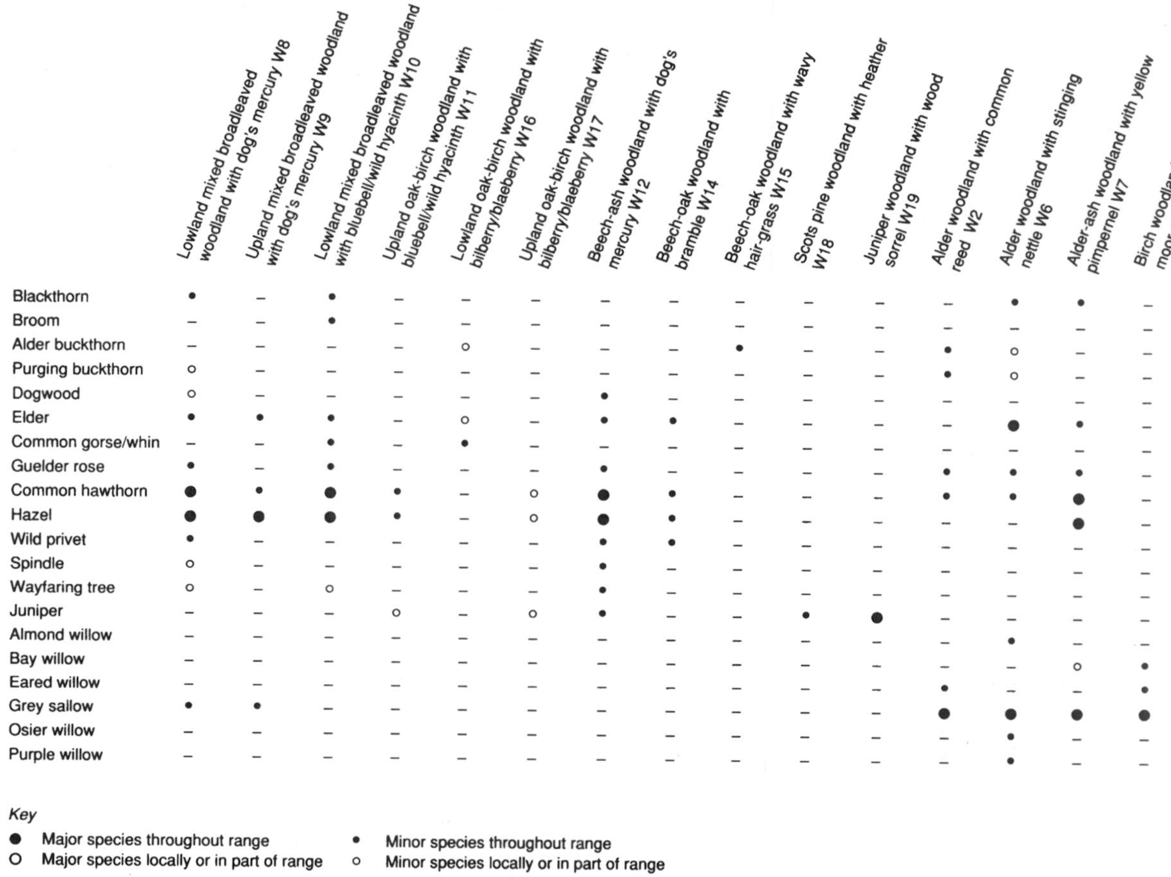

Part 2

Species (rows):
Blackthorn
Broom
Alder buckthorn
Purging buckthorn
Dogwood
Elder
Common gorse/whin
Guelder rose
Common hawthorn
Hazel
Wild privet
Spindle
Wayfaring tree
Juniper
Almond willow
Bay willow
Eared willow
Grey sallow
Osier willow
Purple willow

Woodland types (columns):
Birch woodland with purple moor-grass W4
Alder-ash woodland with yellow pimpernel W7
Alder woodland with stinging nettle W6
Alder woodland with common reed W2
Juniper woodland with wood sorrel W19
Scots pine woodland with heather W18
Beech-oak woodland with wavy hair-grass W15
Beech-oak woodland with bramble W14
Beech-ash woodland with dog's mercury W12
Upland oak-birch woodland with bilberry/blaeberry W17
Lowland oak-birch woodland with bilberry/blaeberry W16
Upland oak-birch woodland with bluebell/wild hyacinth W11
Lowland mixed broadleaved woodland with bluebell/wild hyacinth W10
Upland mixed broadleaved woodland with dog's mercury W9
Lowland mixed broadleaved woodland with dog's mercury W8

Key

● Major species throughout range
○ Major species locally or in part of range
• Minor species throughout range
o Minor species locally or in part of range

Figure 5.2. (continued)

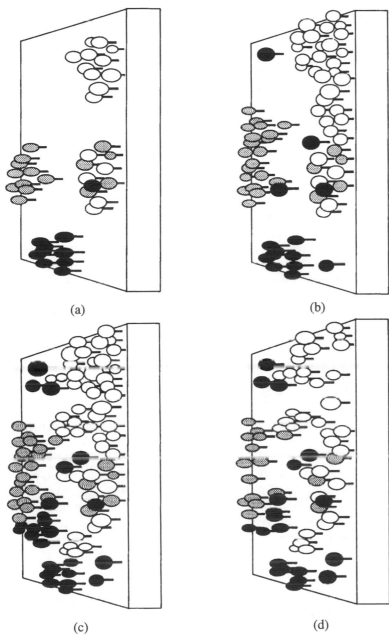

(a)   (b)

(c)   (d)

Figure 5.3. Varied composition and spacing to encourage diversity in new plantings using (a) pure clumps or clumps with two or three well-matched species; (b) varied clump size; (c) varied spacing between clumps; (d) varied spacing within clumps. Reproduced from J. Rodwell and G. Patterson (1994), *Creating New Native Woodlands*. Forestry Commission Bulletin 112, by permission of HMSO

Ancient woodland indicators and naturalised species

There might also be other reasons, not evident from particular classifications themselves, why certain species of tree or shrub found in existing stands should be excluded from planting mixtures or used with different frequency to that observed at present. Past planting, for example, has already obscured some distributions of native tree species that are of intrinsic importance and which ought not to be further confused without careful records being kept of new plantings. Such species might include small-leaved lime (*Tilia cordata*), large-leaved lime (*Tilia platyphyllos*), wild service tree (*Sorbus torminalis*), midland hawthorn (*Crateaegus laevigata*) and the rare whitebeams of the *Sorbus aria* group (Soutar and Peterken, 1989a,b; see Spencer, Chapter 1). As so-called 'ancient woodland indicators', there can be a voguish enthusiasm for planting such trees to give a measure of authenticity to new woods. With other native trees, like the suckering elms of the *Ulmus procera* and *carpinifolia* groups, their complex genetics and uncertain vulnerability to Dutch elm disease might also suggest caution about inclusion. Then, there are some introduced trees, notably sycamore (*Acer pseudoplatanus*), which the NVC shows to have found a niche among British woods that is perfectly predictable from its behaviour within its native range elsewhere in Europe, but which it might be thought imprudent to encourage further, at least in new woodlands comprising truly native species. However, where a strictly native woodland is not sought, the use in planting mixtures of naturalised species like *A. pseudoplatanus* or sweet chestnut (*Castanea sativa*) for example, can be guided by classifications of existing stands.

IDENTIFYING OPTIMAL PRECURSOR VEGETATION

As well as predicting mixtures of trees and shrubs suitable for planting on particular sites, classifications can also identify the vascular and non-flowering plants associated with established stands of different kinds of native woodland. Some of these plants may already be present in existing grasslands, heaths, mires or other vegetation types on sites selected for planting on the basis of their habitat characteristics. Vegetation types with such species can be considered as 'optimal precursors', which could help confirm the selection of a particular woodland type as appropriate to the site and perhaps give a head start to the development of a fuller woodland flora.

An example: W10 Quercus-Pteridium-Rubus woodland

For this type of lowland woodland (W10 in the NVC scheme and often characterised by the presence of bluebell, *Hyacinthoides non-scripta*), such precursors include grasslands with plants like false oat grass (*Arrhenatherum elatius*), Yorkshire fog (*Holcus lanatus*), cock's-foot (*Dactylis glomerata*) and hogweed (*Heracleum sphondylium*), which are indicative of suitable soil conditions. Underscrub with foxglove (*Digitalis purpurea*), red campion (*Silene dioica*) and wood sage (*Teucrium scorodonia*) among clumps of bramble (*Rubus fruticosus*), gorse (*Ulex europaeus*) and broom (*Cytisus scoparius*) might be even better as a starting point for planting, being a little further along the line of succession to the selected woodland type.

General considerations

Clearly, with richer assemblages of herbs, precursors of this kind might themselves have very considerable conservation value: diverse mixtures of herbs on flushed ground, for example, would provide an excellent start for establishing particular kinds of wet woodland but are probably greatly prized already. The advantage of using a classification, however, is that it can provide a descriptive framework within which such judgements of relative value of vegetation types can be made.

In contrast to the above situation, where no species characteristic of the desired woodland type are present, this will be a measure of the task involved in developing a full associated flora. Improved pastures, for example, or rank weedy assemblages on abandoned arable land are likely to be prime targets for planting new woodlands in the future but present a somewhat uncongenial starting point on floristic grounds.

ENCOURAGING DESIRED INVADERS

Classifications can also provide lists of vascular and non-flowering plants which are characteristic of particular woodland types but which are unlikely to be present in existing unwooded vegetation, because they characteristically colonise after the establishment of the trees and shrubs, sometimes very slowly.

Ancient woodland indicators

Desired invaders include those herbaceous plants often called 'ancient woodland indicators', although it should be noted that such indicators vary considerably from one woodland type to another and are not always uniformly informative about the age of stands. For the lowland bluebell woodland, such desired invaders include bluebell (*Hyacinthoides non-scripta*) itself, wood anemone (*Anemone nemorosa*), hairy woodrush (*Luzula pilosa*), yellow archangel (*Lamiastrum galeobdolon*), common dog violet (*Viola riviniana*) and some ferns (*Dryopteris filix-mas, D. dilatata*).

Introductions

The appearance and spread of such species within developing woodlands could be the particular focus of monitoring programmes, but it might also be considered appropriate to encourage their colonisation by seeding or planting (see Packham *et al.*, Chapter 9). This can be costly and its success is uncertain but at least a classification can provide information on the ecologically appropriate situations where such expenditure can be targeted.

AN ECOLOGICAL BASIS FOR WOODLAND DESIGN

Classifications of vegetation can also provide information at the landscape scale to inform the design of whole woods or new forests, as well as of particular stands. It is very important to realise that there may be good ecological reasons for encouraging distinctive patterns of different woodland types in close proximity to one another where there is diversity in the terrain (see Bell, Chapter 3).

Woodland pattern

For example, as we have seen, the lowland bluebell woodland is often found on the concave lower slopes of valley sides where brown earths have developed from shales, drift or downwash. Upslope, in such situations, there are often grits or sandstones with rankers or podzols which support lowland bilberry wood with mixtures of *Quercus*, *Betula*, *Ilex aquifolium*, and *Sorbus aucaparia* over an acidophilous field layer with *Vaccinium myrtillus*, wavy hair-grass (*Deschampsia flexuosa*), heath bedstraw (*Galium saxatile*) and tormentil (*Potentilla erecta*) (W16 *Quercus-Betula-Deschampsia* woodland in the NVC). Between the two, if ground-water emerges in a flush as it percolates through the pervious arenaceous rock and meets the impervious shale or superficials, there is characteristically a stand of *Alnus-Fraxinus* woodland with yellow pimpernel (*Lysimachia nemorum*), creeping buttercup (*Ranunculus repens*), meadowsweet (*Filipendula ulmaria*), lady fern (*Athyrium felix-femina*) and soft rush (*Juncus effusus*) (W7 *Alnus-Fraxinus-Lysimachia* woodland). Such patterns as this can therefore be mimicked when planting trees and shrubs over slopes which are at present unwooded but which offer such geological and edaphic templates (Figure 5.4).

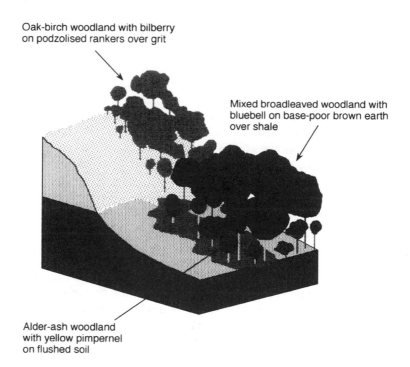

Figure 5.4. Using geological and soil differences to develop mosaics in new native woodlands. Reproduced from J. Rodwell and G. Patterson (1994), *Creating New Native Woodlands,* Forestry Commission Bulletin 112, by permission of HMSO

Open areas

Classifications can also yield information about the characteristic unwooded vegetation associated with surviving stands of different woodland types if, as seems increasingly desirable, new woods are to include open areas (Forestry Commission, 1990). In its characteristic setting on the sandy commons of south-east England, for example, existing stands of the lowland *Vaccinium myrtillus* woodland form part of mosaics with heath vegetation that has ling (*Calluna vulgaris*), bell heather (*Erica cinerea*) and dwarf gorse (*Ulex minor*), calcifugous grassland with sheep's sorrel (*Rumex acetosella*), and scrub with common gorse (*Ulex europaeus*) and *Rubus*. Classifications which include the full realm of variation in such landscapes can provide comparable descriptions of these different elements of the pattern of vegetation and help develop an integrated overview of the potential of a particular stretch of prospective wooded scenery (Rodwell, 1991b, 1992).

SENSITIVE MANAGEMENT OF NEW WOODS

Classifications which are more than just taxonomies of vegetation types can help develop an ecological understanding of the dynamic relationships of woodlands to the climatic and soil factors which determine their composition and range, but also to the various impacts which human activities can have on the developing succession and mature woodland cover (Rodwell and Cooper, 1990). These insights can then help devise sensitive management for the establishment and maintenance of new stands.

Ground preparation

Ground preparation is often desirable to increase aeration and drainage of planting sites but it may disrupt the natural conditions essential for the development of some wetter native woodlands. Or, such operations might encourage the spread of troublesome weeds that could be aggressive competitors to young trees and shrubs. Within the framework of a classification, it is possible to identify such potential aggressors fairly precisely. In the lowland bluebell woodland, for example, *Rubus fruticosus* or, on more free-draining soils, *Pteridium aquilinum* often spread with cultivation and overwhelm the developing field layer. On the more free-draining calcareous soils which support *Fraxinus-Acer campestre* woodland with dog's mercury (*Mercurialis perennis*), by contrast, it is wood false-brome (*Brachypodium sylvaticum*) which commonly becomes abundant after cultivation. Disturbance and compaction, such as ensues with the use of heavy machinery on heavy gleys, can trigger an explosive spread of tufted hair-grass (*Deschampsia cespitosa*) around the margins of *Alnus-Fraxinus-Lysimachia* flushes.

Other influences

Similarly, knowledge about those species which are especially responsive to nutrient enrichment in existing stands of different types of woodland can provide valuable warnings of potential weeds in new woods, where the use of fertilisers is unavoidable to assist establishment. Or again, observations on susceptibility of ground-cover plants to grazing in existing woodlands can inform about likely responses to exclusion of stock and deer from new woods by fencing.

SUMMARY

Classifications of existing woodland vegetation can provide a valuable ecological rationale for creating new woods using native trees and shrubs. The predictive power of such classifications can help select appropriate species and target plantings on sites which already have elements of the appropriate woodland flora. Classifications can also identify desired herbaceous plants whose appearance could be monitored or encouraged and provide guidelines for sensitive management to optimise the development of a full woodland flora. Understanding patterns among existing stands of semi-natural woodlands can also inform the design of new woods where variety among planted mixtures can create ecologically appropriate mosaics.

REFERENCES

Forestry Commission (1990). *Forest Nature Conservation Guidelines*. HMSO, London.
Peterken, G.F. (1981). *Woodland Conservation and Management*. Chapman & Hall, London.
Peterken, G.F. and Game, M. (1984). Historical factors affecting the number and distribution of vascular plant species in the woodlands of central Lincolnshire. *Journal of Ecology*, **72**, 155–182.
Rackham, O. (1980). *Ancient woodland*. Arnold, London.
Rodwell, J.S. (ed.) (1991a). *British Plant Communities*, vol. 1: *Woodlands and Scrub*. Cambridge University Press, Cambridge.
Rodwell, J.S. (ed.) (1991b). *British Plant Communities*, vol. 2: *Mires and Heaths*. Cambridge University Press, Cambridge.
Rodwell, J.S. (ed.) (1992). *British Plant Communities*, vol. 3: *Grasslands and Montane Communities*. Cambridge University Press, Cambridge.
Rodwell, J.S. and Cooper, E.A. (1990). The Development of Vegetation in Recently Established Broadleaved Woodlands in Upland Britain. Unpublished report to the Forestry Commission.
Rodwell, J.S. and Patterson, G. (1994). *Creating New Native Woodlands*. Forestry Commission Bulletin 112. HMSO, London.
Soutar, R. and Peterken, G.F. (1989a). Native trees and shrubs for wildlife. *Tree News*, September, 14–15.
Soutar, R. and Peterken, G.F. (1989b). Regional lists of native trees and shrubs for use in afforestation schemes. *Arboricultural Journal*, **13**, 33–43.

6 Soils and Restoration Ecology

A. J. MOFFAT and G. P. BUCKLEY

INTRODUCTION

Soil is an integral part of the woodland ecosystem, yet it is often neglected or forgotten when planning and planting a woodland on a new site. Instead, reliance may be placed on standard silvicultural practice, often that gained in the course of establishing forests in upland Britain. However, with a desire to increase the amount of broadleaved planting in Britain, coupled with new opportunities in the lowlands (Farm and Community Forestry), it is appropriate to review the need for an understanding of the soil in the woodland ecosystem.

New woodlands are often fashioned today under schemes such as the Community Forests Initiative (Countryside Commission/Forestry Commission, 1989) and the National Forest Strategy (Countryside Commission, 1993), or in rural areas via the Farm Woodland Scheme and its successor, the Farm Woodland Premium Scheme (Ministry of Agriculture, Fisheries and Food, 1992) and the Woodland Grant Scheme (The Forestry Authority, 1994). Available land for planting is of two main types:

- Ex-agricultural land.
- Land disturbed by mineral extraction, industrial development or dereliction.

These forms of land can be very different from those where the majority of woodland is found, or where experience in establishing woodlands has been gained. In some parts of the country, particularly in lowland and southern regions, new farm woodlands will present a woodland cover to some soils that have been in agriculture since woodland clearance began in Neolithic times around 3000 BC (Fowler, 1983; see Bell, Chapter 3). Many disturbed substrates, for example colliery spoil tips, may never have supported a vegetation cover of any description before.

The amount of disturbed land in the UK is sizeable: perhaps over 200 000 hectares. It is composed of land that has been worked for minerals, used for overburden disposal, or left derelict after industrial decline. In some community forest areas, disturbed land represents a significant proportion of potentially available land for tree planting, for example 16% in the Thames Chase area east of London, 16% in the Marston Vale region around Bedford, and 10% in the Forest of Mercia, Leicestershire. The opportunity to plant-up disturbed land with trees is often greater than that for agricultural land, because the former is often in public ownership. The desire to see landscape improvement in these areas is another significant influence on woodland establishment. It is vital, therefore, that the problems posed by these substrates are understood and met.

The Ecology of Woodland Creation. Edited by Richard Ferris-Kaan.

Traditional afforestation has often been accompanied by soil disturbance, amelioration or both. New woodland creation on ex-agricultural and disturbed land may also require soil 'restoration'. An ecological approach can be helpful in this respect. Biotic aspects of soils are considered in Chapter 7 by Harris and Hill. This chapter concentrates instead on the physico-chemical aspects of soils and how these influence potential for new woodland.

SOILS UNDER FOREST

Under natural conditions, what is a forest soil? It is helpful to ask this basic question in order to establish how different or similar are those soils now on offer for woodland establishment. Of course, in the UK with its large range of geological substrates, physiographic and climatic regimes, there are many types of forest soil today. Nevertheless, soils under undisturbed or primary woodland are likely to differ in many respects from those found under agriculture, especially arable husbandry (Peterken, 1993).

Primary woodland

Forest soils tend to exhibit more marked horizonation, have richer organic upper layers, and are more acidic, porous and structured. They are also often thicker, and drier, than their agricultural counterparts. In addition, variability in soil physical and chemical properties differs both vertically and horizontally from that commonly found under agriculture. Soil fauna and microflora, and their functions, are also distinctly different (see Harris and Hill, Chapter 7).

Secondary woodland

Soil properties under secondary woodland will depend on the length of time the site has been under a tree cover and the nature of soil disturbance while under the previous land-use (Peterken, 1993). For example, agricultural land drainage, ploughing, liming and fertiliser application can have marked effects on the soil. Under secondary woodland, some facets of the new soil, such as the pattern of horizontal variability, may take centuries to develop to a stage similar to that found in primary woodland (Wilson, 1992); however, some features, such as soil acidity, can change within decades (e.g. Billett *et al.*, 1988; Moffat and Boswell, 1990).

ESTABLISHING WOODLAND ON EX-AGRICULTURAL SOILS

Distinct differences between 'forest' and 'agricultural' soils therefore exist. Nevertheless, not all such differences are important for the establishment and continued growth of new woodlands. The following features of agricultural soils need to be appreciated.

- *Their texture.* Many recent applications for tree planting grants on farm land have come from areas where clayey soils are predominant, such as Gloucestershire, Hereford and Worcester, Oxfordshire, north-east Wiltshire, Northamptonshire, Norfolk and Suffolk. Clayey soils are often waterlogged in winter months, and many suffer from cracking and drought during the summer (Williamson, 1992).

- *Their base-rich nature.* Agricultural liming has taken place, in some places from Neolithic times. Modern agricultural practice strives to maintain soil pH at 6.5 for arable cropping and 6.0 for grassland (Ministry of Agriculture, Fisheries and Food, 1988). Soil pH will affect many aspects of soil functioning, especially nutrient cycling; in general cycling will be enhanced in soils coming out of agriculture compared to those under moorland or forest.
- *Their higher level of fertility.* Compared with soils traditionally available for afforestation, farmland soils, especially those in the lowlands, are often intrinsically more fertile than heathland or moorland soil types on which the bulk of recent afforestation has taken place; fertility has been augmented by application of fertilisers used to promote arable or grassland husbandry.

Texture

Of itself, a clayey soil texture should not preclude new woodland establishment; forests such as Torridge in Devon, Powerstock in Dorset, Bernwood in Oxfordshire, Rockingham and Kesteven in Northamptonshire were planted on predominantly clayey soils. Nevertheless, winter waterlogging can seriously affect tree survival and early growth; cultivation is often used to locally increase the depth of aerobic soil in which to plant. In dry summers, soil cracking can also reduce tree growth by exposing roots to drought (Moffat *et al.*, 1994); unfortunately, cultivation to alleviate winter waterlogging may exacerbate soil droughtiness in the summer. Apart from shallow cultivation, there is little that can be done in the short term to improve the physical characteristics of clayey soils. Planting success can be maximised by attention to good silviculture, especially in weed control and species choice (Williamson, 1992).

Soil pH

Soil pH is very important to the choice of tree species in a new planting scheme. On ex-agricultural land, soil pH should usually lie between 4 and 8.5; alkaline pH is commonplace, especially on Jurassic and glacial clays in the English Midlands. Most conifers, and many broadleaves, are, in contrast, well adapted to acid soil conditions and grow best in fairly acid media (Pritchett and Fisher, 1987). It is impractical and uneconomic to consider reducing soil pH with artificial additives. Hence, on base-rich soils, species choice will be limited.

Fertility

Perhaps the largest preconception surrounding tree planting on ex-agricultural land is that soils are highly fertile (Potter, 1988; Williamson, 1992). Soil fertility is a desirable property for tree establishment, and it is unusual for tree growth to be affected other than beneficially. However, nutrient imbalance can occur locally, for example if the soil contains an excess of nitrogen over base cations such as calcium or magnesium, or boron. Nutrient deficiency can be induced in these circumstances (Miller and Miller, 1988; Muys, 1990). High NO_3-N production and relative nitrification can also cause stem deformity, without induced deficiencies in other nutrients (Carlyle *et al.*, 1989; Bail and Pederick, 1989). However, some ex-agricultural soils may not be especially fertile, and yields of some tree species, notably of conifers, may be restricted (Fourt *et al.*, 1971; Forestry Commission Research Division Mensuration Branch, 1993). On some sites, fertiliser applications may be beneficial, for example on acidic gleys on the Culm clays in south-west England (Everard, 1974).

Trees planted on ex-agricultural land may also suffer from a lack of mycorrhizal association, which may inhibit performance. Ectomycorrhiza are often inhibited in neutral or alkaline soil (Jackson and Mason, 1984), and even on acidic substrates, inoculation may not take place if the land has been under arable cultivation for many years (Atkinson, 1990). Some tree species are known to have an obligate dependence on mycorrhizal infection, even when soil nutrient levels are adequate. Hence absence of suitable mycorrhiza may predispose some trees to failure. However, no examples of this phenomenon are known in the UK; more research on the place of mycorrhizal inoculation in farm forestry may be warranted. Other features of agricultural soils are unlikely to restrict opportunities for woodland establishment.

Understorey vegetation growth on ex-agricultural land

Soil fertility can have a large indirect effect on tree establishment through the effect it has on weed growth; this can be prolific on fertile soils, and will detrimentally affect survival and growth unless controlled (Williamson, 1992). Rank weed growth will also affect the course of natural colonisation. Although low soil fertility has been regarded almost as a *sine qua non* for successfully creating high species diversity in other habitats such as wild flower meadows, shading by the tree canopy is likely to prove a more important factor controlling the complexity of the vegetation under woodland.

Initially the growth of rank grasses and tall herbs can be expected to delay the establishment of volunteer shrub and woodland field layers, but the National Vegetation Classification (Rodwell, 1991) throws some light on the likely sequence of scrub development. Thus *Crataegus-Hedera* and *Prunus-Rubus* scrub communities (W21,22) may develop on fertile, base-rich soils, with the invasion of shrubs such as hawthorn (*Crateagus monogyna*), blackthorn (*Prunus spinosa*), bramble (*Rubus fruticosus*) and elder (*Sambucus nigra*) over field layers initially dominated by nettle (*Urtica dioica*), cleavers (*Galium aparine*), cow parsley (*Anthriscus sylvestris*), false wood-brome (*Brachypodium sylvaticum*) or ivy (*Hedera helix*).

Other indications of how fertile agricultural soils might determine the development of woodland plant communities can be inferred from case studies of vegetation change in woodlands subject to gradually increasing levels of atmospheric pollution. Some authors consider that changes in the herb layer favouring nitrophilous species such as common hemp-nettle (*Galeopsis tetrahit*), raspberry (*Rubus idaeus*), ground-elder (*Aegopodium podagraria*) and rosebay willowherb (*Epilobium angustifolium*) in forest vegetation may be triggered by atmospheric inputs (van Breeman and van Dijk, 1988; Falkengren-Grerup, 1986). Similarly, Thimonier *et al.* (1992) recorded species changes in Amance Forest, Nancy, France over a 19-year period, and attributed increases in *Galium aparine,* ground ivy (*Glechoma hereracea*) and hogweed (*Heracleum sphondylium*), among other species, to increased usage of nitrogen fertiliser in the agricultural hinterland. Such species are likely to prove persistent in young woods, and it may take several years before the development of forest conditions (in the absence of pollution) leads to the formation of a richer community.

ESTABLISHING TREES ON DISTURBED LAND

Just as agricultural soils differ in some respects from those commonly found under wood-land, so too do the properties of 'urban' soils. Craul (1985, 1992) describes in detail the main properties (Table 6.1). They are discussed briefly below:

Table 6.1. Soil property differences between disturbed and undisturbed land

Compared with undisturbed soils, disturbed soils tend to have:

1. Greater vertical and horizontal variability
2. A modified soil structure, often compact
3. The presence of a surface crust on bare soil that is usually hydrophobic
4. A modified soil reaction, often alkaline (pH>7)
5. Restricted aeration and water drainage
6. Interrupted nutrient cycling, and a modified soil organism population and activity
7. The presence of anthropogenic materials and other contaminants
8. Highly modified soil temperature regimes

- Soil materials on disturbed land are either degraded or simply derived from mineral rock or overburden. Hence, infertility is commonplace, especially of nitrogen and phosphorus. Deficiencies in other nutrients can occur locally.
- The amount of rootable materials can be of limited availability, and may be too small to sustain the growth of a tree crop, supplying nutrients, moisture and anchorage.
- Soil materials usually have large bulk densities, having been compacted by movement or vehicle traffic. Compact soils prevent tree root extension, and encourage drought and premature windblow.
- Disturbed soils often have alkaline reaction, contaminated by lime-rich materials such as cement. Alkaline soils will reduce choice of suitable tree species.
- Soil aeration and drainage are often impaired on disturbed sites, in some cases due simply to the large soil bulk density occurring, but sometimes a consequence of shallow site gradient, adverse soil texture, or both. Unless resolved, waterlogging and anaerobic conditions can seriously restrict tree species choice to tolerant species such as willows (*Salix* spp.), alders (*Alnus* spp.) and some poplars (*Populus* spp.); few species can tolerate permanent waterlogging close to the ground surface (Gill, 1970).
- New materials used for tree planting, for example mineral spoils, may contain little or no organic matter. They are therefore comparatively sterile, and only small populations of soil fauna and microflora may be present. Soil nutrient cycling will therefore be inhibited, and the long-term survival of trees compromised.
- Some materials on disturbed sites may be toxic (Bridges, 1987). Depending on the severity of the toxicity, trees may not grow at all, or may survive in stunted form, akin to growth on natural metalliferous substrates such as serpentine (Brooks, 1987).

Tree growth potential on disturbed land

Relatively little data is available on the productivity of trees on derelict and wasteland sites. Existing information has been based on yield class data obtained from standard top height/age curves published by the Forestry Commission (Edwards and Christie, 1981), although the crops concerned were often rather too young for reliable forecasts to be made, and were planted on infertile or toxic sites where growth might be expected to be limited in the long term. However, Broad (1979) showed that in Wales, the yields of pine (*Pinus* spp.) and larch (*Larix* spp.) growing on undisturbed but weathered colliery spoil (8–14 m³ ha⁻¹ yr⁻¹ corresponding to 3.6–5.4 t ha⁻¹ yr⁻¹ of dry matter) were only slightly less than similar crops on adjacent upland brown earth soils. Moderate yields were also found on slate wastes, furnace ash and restored sand quarries, while relatively poor growth occurred on recently landscaped opencast coal sites with thin soil cover (Dennington and Chadwick, 1981). However, recent research (Bending *et al.*, 1991) shows that even on this substrate, satisfactory yields are possible, if not frequently obtained. Table 6.2 summarises several studies of timber yield from stands on restored disturbed land.

Table 6.2. Yields of woody species on restored derelict land

Type of site	Species	Yield (dt ha⁻¹yr⁻¹)
Opencast coal sites	Lodgepole pine	2.7
Colliery spoil	Common alder	0.3–0.9
	Corsican pine	5.4
	False acacia	0.8
	Grey alder	1.0
	Japanese larch	3.6
	Lodgepole pine	4.5
	Sallow	1.3
	Scots pine	4.1
	Silver birch	0.1–1.0
	White poplar	0.7
Pulverised fuel ash	Poplar 'robusta'	0–3
	Silver birch	0–3
	Sitka spruce	2–5
	Sycamore	0–3
Slate waste	Corsican pine	5.4
	Japanese larch	3.6–5.4
	Lodgepole pine	2.7
	Noble fir	5.4–7.2
	Scots pine	3.2–6.3
	Western hemlock	5.4
Sandstone waste	Lodgepole pine	3.2
Limestone waste	Scots pine	4.5
Silica waste	Corsican pine	0.9–2.7 2.7
Furnace ash	Common alder	3.5
	Corsican pine	5.4
	Japanese larch	3.2
	Lodgepole pine	3.6
	Silver birch	2.5
	Sitka spruce	2.7

Dennington *et al.* (1983) examined the potential of different tree species for energy coppice on disturbed land. Their results showed dry matter production for trees aged 9 to 20 years of between 0.1–7.2 t yr^{-1}, depending on the limitations of the site, with the majority giving equivalent mean annual production of 4–5 m^3 ha^{-1} yr^{-1}. The overall mean yield, 3.3 t ha^{-1} yr^{-1}, is considerably smaller than yields of around 10 t ha^{-1} yr^{-1} obtained for short rotation coppice grown at close (1 × 1m) spacing (Tabbush, 1993), but similar to production of 2.5–6 t ha^{-1} yr^{-1} for traditional native broadleaved coppice (Begley and Coates, 1961).

In summary, it appears that although some types of disturbed land are likely to adversely affect tree yield, on most sites plantations can be established that will meet landscape objectives, and may still provide modest timber production over time. Some sites, if reclaimed well, may succeed in achieving levels of productivity found in normal forest crops.

REMEDYING THE ADVERSE CHARACTERISTICS OF DISTURBED LAND

Ecologists have long studied the natural colonisation of disturbed and newly formed substrates. From their work, principles for the reclamation of such substrates to 'speed up' revegetation have evolved (Bradshaw, 1983). Foresters, too, have been heavily involved in the afforestation of sites suffering in similar ways to disturbed ones. Their techniques can often be applied in these circumstances (Wilson, 1985). The following four sections describe how recent research findings from woodlands on semi-natural sites have been used to formulate modern standards of reclamation practice in Great Britain (Moffat and McNeill, 1994).

Soil fertility

Raw, new substrates are usually infertile. Many ecological studies demonstrate that fertility is built up over time, for example the classic studies on glacial moraine sequences in America (Crocker and Major, 1955; Crocker and Dickson, 1957). In the UK, most studies of this kind have taken place on new, man-made, substrates. Early studies on colliery spoil tips include those by Hall (1957) and Crampton (1967). The most important research project was undertaken in the 1970s and 1980s by A.D. Bradshaw and colleagues on china clay spoil in Cornwall (see Bradshaw (1983) for summary of this work). They were able to demonstrate increasing fertility associated with organic matter accumulation, and related this fertility trend to changes in vegetation composition, just as Crocker and colleagues had done in the USA. They proposed that ecosystem nitrogen content was very important; a self-sustaining ecosystem was dependent upon a capital of *c.*1000 kg N ha^{-1}. Bradshaw (1983) went on to propose that this amount of nitrogen was necessary to support a range of ecosystems.

An amount of 800–1000 kg soil N ha^{-1} to support a woodland cover has recently been endorsed by Kendle and Bradshaw (1992), though Finegan (1984) argued that its applicability for trees remained to be substantiated. Evaluation is inevitably prone to difficulty because tree growth depends not on site *capital* but *supply*, which may vary considerably from region to region. There are many examples of mineral sites where tree growth is sustained on capitals considerably less than the level advocated by Bradshaw (e.g. Gadgil, 1971; Humphries and Guarino, 1987). Even on china clay spoil, trees have been established on a site with <600 kg soil N ha^{-1} (Moffat, unpublished). Nevertheless, on many sites, nitrogen deficiency may well limit tree species choice and general tree growth unless remedial action is taken. For example, nitrogen-fixing species like alders (*Alnus* spp.) can be included in the species composition; alternatively, the site can be improved using sewage sludge or other organic wastes (Moffat and McNeill, 1994).

Using these simple restoration strategies, nitrogen levels in the soil ecosystem can be raised rapidly. It is thus eminently possible to establish woodland on comparatively infertile substrates. Alternatively, it may be possible to propagate directly from tolerant ecotypes already growing on disturbed land, as Good *et al.* (1985) found in willow (*Salix* spp.) and birch (*Betula* spp.) populations on opencast coal spoils in south Wales.

Soil thickness

On ex-agricultural land, soil thickness is rarely limiting for tree growth, but on man-made sites it may be necessary to import soil materials in which to establish the new woodland. How much is required, not just to establish growth in early years, but to promote development to maturity? Ecological studies of tree rooting depth in natural soils can help to determine soil requirements (e.g. Perry, 1982; Sutton, 1991). So, too, can recent research on the water needs of mature tree crops.

Tree rooting depth appears to be determined by a number of factors including tree species, soil physical strength, and degree of waterlogging and anaerobism. In loose, well-aerated soils, there appears to be little difference between species in potential rooting depth (Bibelriether, 1966). Instead, the ability of different species to tolerate adverse soil conditions is the principal reason why different trees exhibit contrasting root systems. Most temperate trees have root systems in the range of 1–1.5 m deep. Information gathered from a survey of the root plates of *c.* 3500 trees (Gasson and Cutler, 1990) shows that 44% were less than 1 m deep, and 95% were less than 2 m deep (Table 6.3).

Restoring man-made sites with a soil depth of 1–2 m is suggested by the kind of research summarised above. Nevertheless, soil depth should also be determined by reference to its texture (particle size distribution) and stoniness. Sandy and stony soils hold less water available for tree uptake during the growing season. Hence, proportionately thicker amounts of these soils are necessary to supply the water needs of mature trees compared with loamy, stoneless soils or both. Moffat (1995) has modelled soil thickness taking into account these water requirements. Figures 6.1 to 6.4 show soil thickness requirements for various soil textures and stone contents in England and Wales. Change across the country is related to the variation in summer rainfall. It is clear from the modelling exercise that in wetter parts of the country, soil depths of 0.5 m may be sufficient to keep trees supplied with water in the summer, but in the drier south-east, and especially on stony, sandy soils, thicknesses in excess of 2 m may be required.

Soil physical conditions

Soil physical conditions are rarely in need of amelioration before tree planting. Only if a plough pan is present is root penetration likely to be impeded. However, on other types of land, soil disturbance usually leads to soil compaction and consequent need for restoration.

Table 6.3. Summary of data from the Kew Wind Blown Tree Survey on maximum depths of roots in the root plates of windblown trees. The number of trees having maximum root plate depths in the categories <0.5 m, 0.5–1.0 m, 1.0–1.5 m, 1.5–2.0 m and >2.0 m are shown. Adapted from P.E. Gasson and D.F. Cutler, 1990, *Arboricultural Journal,* **14**, 193–264, and reproduced from M.C. Dobson and A.J. Moffat (1993), *The Potential for Woodland Establishment on Landfill Sites,* Department of the Environment. Crown copyright is reproduced with the permission of the Controller of HMSO

Genus	Maximum root plate depth (m)					Total number of trees	Range of root plate depth (m)
	<0.5	0.5–1.0	1.0–1.5	1.5–2.0	>2.0		
Apple	1	2	4	0	1	8	0.45–2.70
Ash	0	10	14	4	3	31	0.75–2.80
Beech	5	28	52	14	4	103	0.10–2.80
Birch	4	13	13	1	1	32	0.10–3.00
Cedar	1	1	1	0	0	3	1.00–2.00
Cherry	0	0	3	3	0	6	1.00–1.55
Chestnut	1	6	14	1	2	24	0.20–2.19
Cypress	3	9	2	0	0	14	0.65–1.81
Douglas fir	1	3	1	0	0	5	0.30–1.45
False acacia	3	1	0	0	0	4	1.59–2.00
False cypress	2	3	2	1	0	8	0.85–1.30
Fir	1	4	4	3	2	14	0.25–2.17
Hawthorn	2	1	0	0	0	3	0.40–0.80
Hazel	1	1	0	0	0	2	0.35–0.75
Hickory	1	1	2	0	0	4	0.94–1.94
Holly	1	12	1	0	0	14	0.33–1.00
Honey locust	2	1	1	0	0	4	0.50–1.72
Hornbeam	0	12	7	0	1	20	0.50–2.10
Horse chestnut	0	4	2	0	0	6	0.50–1.40
Indian bean tree	4	1	0	0	0	5	0.62–1.21
Larch	1	8	11	3	1	24	0.30–2.20
Lime	1	6	12	4	3	26	0.12–2.60
Maple	0	15	14	2	0	31	0.50–1.82
Mulberry	1	1	1	0	0	3	0.81–1.50
Oak	4	39	62	31	9	145	0.30–2.05
Pine	2	8	16	5	1	32	0.40–3.00
Plane	2	1	0	0	0	3	0.80–1.00
Poplar	0	2	3	6	2	13	0.80–2.43
Rowan	3	4	3	0	0	10	0.40–1.35
Southern beech	2	1	6	1	0	10	0.33–1.58
Spruce	3	21	10	1	1	36	0.30–2.14
Tulip tree	1	0	1	2	0	4	0.93–2.00
Walnut	1	3	0	0	1	5	0.30–2.14
Willow	2	1	4	0	0	7	0.20–1.22
Yew	4	1	1	0	0	6	0.50–1.70
Others	4	19	6	3	0	32	0.30–1.75
Total	64	243	273	85	32	697	0.10–3.00
Percentage	9.2	34.9	39.2	12.2	4.6	100	

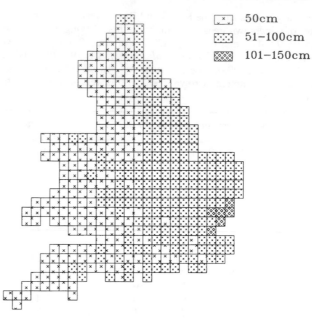

Figure 6.1. Soil thickness needed to satisfy the water requirements of a mature tree crop on stoneless, loamy and clayey soils

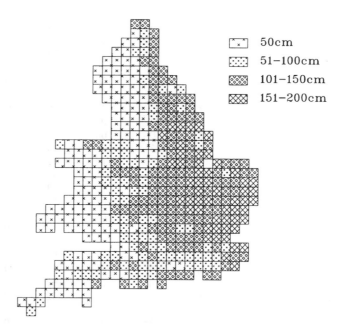

Figure 6.2. Soil thickness needed to satisfy the water requirements of a mature tree crop on stony (30% by volume), loamy and clayey soils

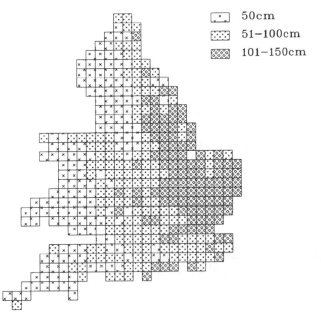

Figure 6.3. Soil thickness needed to satisfy the water requirements of a mature tree crop on stoneless sandy soil

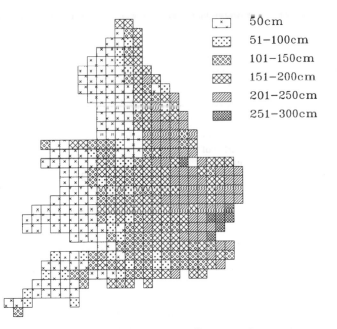

Figure 6.4. Soil thickness needed to satisfy the water requirements of a mature tree crop on a stony (30% by volume) sandy soil

How tolerant are trees to compaction? Studies of natural tree rooting can help to provide the answer. Tree species vary in their ability to cope with compact substrates. For example, Bibelriether (1966) ranked mechanical root penetration for a range of common tree species (Table 6.4), from many observations of rooting habit. Nevertheless, constraints on root growth posed by the soil environment are probably of greater importance (Röhrig, 1966). Features of the soil which hinder root development include mechanical resistance, poor soil aeration, and waterlogging. Growth tends to decrease linearly with increasing soil density (Dobson and Moffat, 1993). Soil densities greater than about 1.6 to 1.8 g cm^{-3} effectively prevent root growth; soils of this density require amelioration, usually by means of cultivation. Deep ripping or subsoiling during summer months when the soil is at its driest is the most effective way (Moffat and McNeill, 1994). Figure 6.5 shows the effect of ripping on soil bulk density at a restored sand and gravel quarry in Hampshire, England.

Table 6.4. Characteristics of rooting ability for a range of tree species. Adapted from H. Bibelriether, 1966, *Allgemeine Forst und Zeitschrift*, **21**, 805–818, and reproduced from M.C. Dobson and A.J. Moffat (1993), *The Potential for Woodland Establishment on Landfill Sites*, Department of the Environment. Crown copyright is reproduced with the permission of the Controller of HMSO

Species	Typical type of root system	Typical root depth (m)	Mechanical root penetration
Ash	Surface root	1.1	Medium
Aspen	Surface root	1.3	High
Birch	Heart root	1.8	Medium
Beech	Heart root	1.3	Low
Common alder	Heart/surface root	2.0	High
Corsican pine	Tap root	–	Medium
Douglas fir	Heart root	2.0	High
English oak	Tap root	1.5	High
European larch	Heart root	2.0	High
Hornbeam	Heart root	1.6	Medium
Japanese larch	Heart root	–	Medium
Lime	Heart root	1.3	Low
Norway maple	Heart root	1.0	–
Norway spruce	Surface root	2.0	Low
Red oak	Heart root	1.6	Medium
Scots pine	Tap root	2.1	High
Sessile oak	Tap root	1.5	High
Silver fir	Tap root	2.0	High
Sycamore	Heart root	1.3	Low
White pine	Surface root	1.7	Low

Tolerance of toxicity

Compared with undisturbed soils, soil materials on disturbed sites can often be chemically hostile for plant growth. Some substrates may be intrinsically toxic, for example colliery spoils containing iron pyrites (FeS_2), which may weather to produce intensely acid conditions. Others may be contaminated, containing foreign introduced materials, for example heavy metals such as copper and lead.

Many tree species have been established on moderately acid spoils, able to tolerate conditions as acid as pH 3.5. Some species, notably *Alnus glutinosa*, *Betula pendula* and

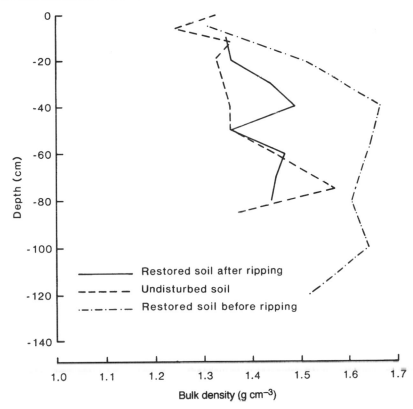

Figure 6.5. The effect on soil bulk density of deep ripping using 75 cm winged tines at a restored sand and gravel quarry in Hampshire, England

false acacia (*Robinia pseudoacacia*) are capable of extending roots into even more acidic substrates (Jobling and Stevens, 1980). The acid regenerating potential of these substrates can be countered by liming (Costigan *et al.*, 1981), and this will increase tree productivity. However, there are likely to be other ecological consequences, for example in restricting the development of acid-tolerant vegetation and effectively reducing site variability.

The tolerance of trees to toxic elements such as heavy metals is quite controversial. Eltrop *et al.* (1991) suggested that woody plant species seem to have a lower degree of tolerance than grasses and other herbaceous species. On heavily contaminated substrates, this seems to be true, yet Zöttl (1985) has shown that some tree species are able to take up large amounts of metals without restricting growth. The existence of specific heavy-metal-tolerant races in trees has yet to be conclusively demonstrated (Bradshaw, 1981), although there are indications that distinct ecotypes do occur in some species exposed to large concentrations of some metals (Eltrop *et al.*, 1991). However, stunted tree form is often a sign of intolerance to native metal concentrations.

On sites severely affected by contamination, some form of restoration is required to allow sustained tree growth into maturity. Methods are discussed by Bridges (1991). They include stabilising the pollutant *in situ*, or containing it physically with barriers and covering systems. It is important that a suitable thickness of soil is provided above any cap over the contaminated materials.

CONCLUSIONS

Soil is a vital part of the forest ecosystem (Moffat, 1991). An understanding of the soil on land where tree planting is proposed is valuable, and invaluable if the soil has suffered disturbance through industrial activity. Soil 'restoration' may be required on the latter substrate to facilitate tree survival and woodland development.

REFERENCES

Atkinson, D. (1990). Farm-forest systems: some implications for nutrient conservation, soil conditions and root development. *SEESOIL*, **6**, 50–60.

Bail, I.R. and Pederick, L.A. (1989). Stem deformity in *Pinus radiata* on highly fertile sites: expression and genetic variation. *Australian Forestry*, **52**, 309–320.

Begley, C.D. and Coates, A.E. (1961). *Estimating Yield of Hardwood Coppice for Pulpwood Growing*. Forestry Commission Report on Forest Research 1959/60. HMSO, London, 189–196.

Bending, N.A.D., Moffat, A.J. and Roberts, C.J. (1991). Site factors affecting tree response on restored opencast ground in the south Wales coalfield. In: Davies, M.C.R. (ed.). *Land Reclamation. An End to Dereliction?* Elsevier Applied Science, London, 347–356.

Bibelriether, H. (1966). Root development of some tree species in relation to soil properties. *Allgemeine Forst und Zeitschrift*, **21**, 805–818.

Billett, M.F., Fitzpatrick, E.A. and Cresser, M.S. (1988). Long-term changes in the acidity of forest soils in north-east Scotland. *Soil Use and Management*, **4**, 102–107.

Bradshaw, A.D. (1981). Growing trees in difficult environments. In: *Research for Practical Arboriculture*. Forestry Commission Occasional Paper 10. Forestry Commission, Edinburgh, 93–106.

Bradshaw, A.D. (1983). The reconstruction of ecosystems. *Journal of Applied Ecology*, **20**, 1–17.

Bridges, E.M. (1987). *Surveying Derelict Land*. Clarendon Press, Oxford.

Bridges, E.M. (1991). Dealing with contaminated soils. *Soil Use and Management*, **7**, 151–158.

Broad, K.F. (1979). *Tree Planting on Man-made Sites in Wales*. Forestry Commission, Edinburgh.

Brooks, R.R. (1987). *Serpentine and its Vegetation*. Croom Helm, London.

Carlyle, J.C., Turvey, N.D., Hopmans, P. and Downs, G.M. (1989). Stem deformation in *Pinus radiata* associated with previous land use. *Canadian Journal of Forest Research*, **19**, 96–105.

Costigan, P.A., Bradshaw, A.D. and Gemmell, R.P. (1981). The reclamation of acidic colliery spoil. I. Acid production potential. *Journal of Applied Ecology*, **18**, 865–878.

Countryside Commission (1993). *The National Forest Strategy*. Countryside Commission Paper 411. Countryside Commission, Cheltenham.

Countryside Commission/Forestry Commission (1989). *Forests for the Community*. Countryside Commission Paper 270. Countryside Commission, Cheltenham.

Crampton, C.B. (1967). Soil development on tips in South Wales. *Sylva*, **47**, 12–14.

Craul, P.J. (1985). A description of urban soils and their desired characteristics. *Journal of Arboriculture*, **11**, 330–339.

Craul, P.J. (1992). *Urban Soil in Landscape Design*. Wiley, New York.

Crocker, R.L. and Major, J. (1955). Soil development in relation to vegetation and surface age at Glacier Bay, Alaska. *Journal of Ecology*, **43**, 427–449.

Crocker, R.L. and Dickson, B.A. (1957). Soil development on the recessional moraines of the Herbert and Mendenhall glaciers, south-eastern Alaska. *Journal of Ecology*, **45**, 169–185.

Dennington, V.N. and Chadwick, M.J. (1981). *An Assessment of the Potential of Derelict and Industrial Wasteland for the Growth of Energy Crops: Yield Assessments and Management Strategies*. Energy Technology Support Unit, Department of Energy.

Dennington, V.N., Chadwick, M.J. and Chase, D.S. (1983). Energy cropping on derelict and waste land. *Journal of Environmental Management*, 16, 241–260.

Dobson, M.C. and Moffat, A.J. (1993). *The Potential for Woodland Establishment on Landfill Sites*. HMSO, London.

Edwards, P.N. and Christie, J.M. (1981). *Yield Models for Forest Management*. Forestry Commission Booklet 48. Forestry Commission, Edinburgh.

Eltrop, L., Brown, G., Joachim, O. and Brinkman, K. (1991). Lead tolerance of *Betula* and *Salix* in the mining area of Mechernich, Germany. *Plant and Soil*, 131, 275–285.

Everard, J.E. (1974). *Fertilisers in the Establishment of Conifers in Wales and Southern England*. Forestry Commission Booklet 41. HMSO, London.

Finegan, B.G. (1984). Forest Succession. *Nature*, 312, 109–114.

Forestry Authority (1994). *Woodland Grant Scheme*. Forestry Authority, Edinburgh.

Falkengren-Grerup, U. (1986) Soil acidification and vegetation changes in deciduous forest in southern Sweden. *Oecologia (Berlin)*, 70, 339–347.

Forestry Commission Research Division Mensuration Branch (1993). Site Classification and Yield Prediction for Lowland Sites in England and Wales. Forestry Commission and Ministry of Agriculture, Fisheries and Food Joint Contract CSA 1563 Final Report, Forestry Commission, Farnham.

Fourt, D.F., Donald, D.G.M., Jeffers, J.N.R. and Binns, W.O. (1971). Corsican pine (*Pinus nigra* var. *maritima* (Ait.) Melville) in Southern Britain. A study of growth and site factors. *Forestry*, 44, 189–207.

Fowler, P.J. (1983). *The Farming of Prehistoric Britain*. Cambridge University Press, Cambridge.

Gadgil, R.L. (1971). The nutritional role of *Lupinus arboreus* in coastal sand dune forestry, I. The potential influence of undamaged lupin plants and nitrogen uptake by *Pinus radiata*. *Plant and Soil*, 34, 357-367.

Gasson, P.E. and Cutler, D.F. (1990). Tree root plate morphology. *Arboricultural Journal*, 14, 193–264.

Gill, C.J. (1970). The flooding tolerance of woody species – a review. *Forestry Abstracts*, 31, 671–688.

Good, J.E.G., Williams, T.G. and Moss, D. (1985). Survival and growth of selected clones of birch and willow on restored opencast sites. *Journal of Applied Ecology*, 22, 995–1008.

Hall, I.G. (1957). The ecology of disused pit heaps in England. *Journal of Ecology*, 45, 689–720.

Humphries, R.N. and Guarino, L. (1987). Soil nitrogen and the growth of birch and buddleia in abandoned chalk quarries *Reclamation and Revegetation Research*, 6, 55–61

Jackson, R.M. and Mason, P.A. (1984). *Mycorrhiza*. Edward Arnold, London.

Jobling, J. and Stevens, F.R.W. (1980). *Establishment of Trees on Regraded Colliery Spoil Heaps*. Occasional Paper 7. Forestry Commission, Edinburgh.

Kendle, A.D. and Bradshaw, A.D. (1992). The role of soil nitrogen in the growth of trees on derelict land. *Arboricultural Journal*, 16, 103–122.

Miller, H.G. and Miller, J.D. (1988). Response to heavy nitrogen applications in fertilizer experiments in British forests. *Environmental Pollution*, 54, 219–231.

Ministry of Agriculture, Fisheries and Food (1988). *Fertiliser Recommendations*. MAFF Reference Book 209. HMSO, London.

Ministry of Agriculture, Fisheries and Food (1992). *The Farm Woodland Premium Scheme. Rules and Procedures*. MAFF PB 0825, London.

Moffat, A.J. (1991). The importance of soils in modern forestry. *Forestry*, 64, 217–238.

Moffat, A.J. (1995). Minimum soil depths for the establishment of woodland on disturbed ground. *Arboricultural Journal*, 19.

Moffat, A.J. and Boswell, R.C. (1990). Effect of tree species and species mixtures on soil properties at Gisburn Forest, Yorkshire. *Soil Use and Management*, 6, 46–51.

Moffat, A.J., Armstrong, A.T. and Collyer, E.L. (1994). Site preparation for tree establishment on lowland clay soils. *Quarterly Journal of Forestry*, 88, 35–41.

Moffat, A.J. and McNeill, J.D. (1994). *Restoring Disturbed Land for Forestry*. Forestry Commission Bulletin 110. HMSO, London.

Muys, B. (1990). N-excess in the forest: effects and possible measures. *Silva Gandavensis*, 55, 35–42.

Perry, T.O. (1982). The ecology of tree roots and the practical significance thereof. *Journal of Arboriculture*, **8**, 197–211.

Peterken, G.F. (1993). *Woodland Conservation and Management,* 2nd edn. Chapman and Hall, London.

Potter, M.J. (1988). Establishing new farm woodland. In: Hibberd, B.G. (ed.). *Farm Woodland Practice*. Forestry Commission Handbook 3. HMSO, London, 30–37.

Pritchett, W.L. and Fisher, R.F. (1987). *Properties and Management of Forest Soils*, 2nd edn. Wiley, New York.

Rodwell, J.S. (ed.) (1991). *British Plant Communities*, vol. 1: *Woodlands and Scrub*. Cambridge University Press, Cambridge.

Röhrig, R. (1966). Root development of forest trees in relation to ecological conditions. Part I and II. *Forstarchiv*, **37**, 217–229 and 237–249. Translated from the German by the Canadian Department of Forestry and Rural Development. Translation No. 101.

Sutton, R.F. (1991). Soil properties and root development in forest trees: a review. Forestry Canada, Ontario Region, Information Report O-X-413.

Tabbush, P.M. (1993). *Coppiced Trees as Energy Crops*. ETSU Report B 1291. Energy Technology Support Unit, Harwell.

The Woodland Grant Scheme Applicant's Pack (1994). The Forestry Authority, Edinburgh.

Thimonier, A., Dupouey, J.L. and Timbal, J. (1992). Floristic changes in the herb-layer of a deciduous forest in the Lorraine Plain under the influence of atmospheric pollution. *Forest Ecology and Management*, **55**, 149–167.

van Breeman, N. and van Dijk, H.F.G. (1988). Ecosystem effects of deposition of nitrogen in the Netherlands. *Environmental Pollution*, **54**, 249–274.

Williamson, D.R. (1992). *Establishing Farm Woodlands*. Forestry Commission Handbook 8. HMSO, London.

Wilson, B.R. (1992). The Nature and Pattern of Soils under Ancient Woodland in Southern England. Unpublished PhD thesis, University of Reading.

Wilson, K. (1985). *A Guide to the Reclamation of Mineral Workings for Forestry*. Forestry Commission Research and Development Paper 141. Forestry Commission, Edinburgh.

Zöttl, H.W. (1985). Heavy metal levels and cycling in forest ecosystems. *Experientia*, **41**, 1104–1113.

7 Soil Biotic Communities and New Woodland

J. A. HARRIS and T.C.J. HILL

INTRODUCTION

When considering the re-establishment of woodlands on degraded sites, considerable attention must be paid to the soil on which the woodland will exist. The soil will shape, and in turn be shaped by, the interaction of abiotic and biotic factors. Without the biological component soils can simply not exist, as without organic matter the mineral particles will be washed or blown away. Furthermore the majority of soils have formed over millennia, and even today soils as old as 10 000 years can be found, still functioning as a means of support and nutrient supply for the plant and animal communities growing on them.

It can also be shown that the measurement and understanding of soil microbial processes is central to establishing a self-sustaining woodland ecosystem, and may be used to measure the success of restoration. This understanding may be employed to make informed interventions in terms of organic and inorganic amendments, cultivation and biological management.

THE ROLE OF LIVING ORGANISMS IN SOIL FORMATION AND FUNCTION

Soil formation

Pioneer colonisers of consolidated (rock) materials are principally lichens. These produce acids which greatly increase the rate of formation of particulate materials, by accelerating chemical weathering. Colonisers of loose materials such as sand are usually more complex plants, although in very moist sites cyanobacteria can play an important role in fixing nitrogen into an organic form. Their roots bind soils together, preventing erosion, and can force open cracks in solid materials by hydrostatic pressure, and provide channels for aeration and drainage when the plant dies. They are the primary producers of the system and give rise, in one way or another, to most of the organic material in the system by litter fall and incorporation. The total amount added per annum will vary from habitat to habitat, and from one year to another (Table 7.1). At the same time plants extract water and mineral nutrients from the soil, and also act as a physical mechanism of support.

The Ecology of Woodland Creation. Edited by Richard Ferris-Kaan.

Table 7.1. Annual inputs of organic matter to soil

Habitat	Organic matter input (tonne ha^{-1} year^{-1})	Soil organic matter content (%)
Tropical forest	25	<5
Prairie	5	15
Pine forest	2.5	>90

Some vertebrates burrow in the soil to obtain shelter, bringing about mixing of layers, although this is far less significant than that brought about by invertebrates. Earthworms, nematodes, mites, springtails, millipedes, gastropods and numerous insect groups are responsible in many cases for the initial stages of plant litter breakdown, and tend to be concentrated in the top 5 cm of the soil. Earthworms are probably the most important members of this group, certainly in temperate soil systems, as they play a crucial role in the development of a stable soil structure. They break up organic matter and excrete faecal material, which includes clay minerals, giving rise to water stable aggregates. This is essential to the maintenance of an open, freely draining soil structure.

The major groups of soil micro-organisms (bacteria, fungi, actinomycetes, algae and protozoa) are responsible for mediating many plant nutrient transformations and also play a part in the maintenance of soil structure. Bacteria excrete polysaccharides which bind mineral particles together, while fungi enmesh soil particles in their hyphae.

Organic matter is in many respects the most important component of the soil, as it is responsible for turning a loose collection of mineral particles into a soil. It acts as a buffer against environmental stresses and disturbances and contributes to the structural stability of the soil.

Soil function

Micro-organisms make a vital contribution to soil function. Many of the transformations which occur during cycling of nutrients are dependent on the soil microbial community, e.g. the carbon cycle (Figure 7.1). Without the microbial community dead organic matter would not be recycled to the atmosphere, and humic material responsible for long-term stabilisation of the soil structure would not be formed.

The microbial community is involved in the transformations of all mineral elements, but it is worth noting its central role in the nitrogen cycle (Figure 7.2). Microbes are responsible for: *nitrogen fixation*, in association with higher plants and as free living organisms; *ammonification* (the conversion of organic nitrogen to plant available inorganic forms); *nitrification* (the conversion of ammonium to nitrates) and *denitrification* (the production of dinitrogen from inorganic forms for recycling to the atmosphere).

DISRUPTION OF SOIL FUNCTION

Disturbance events may be broadly categorised as:

● *Destructive disturbance*, in which part of the existing biomass is killed, thus providing nutrients for the part that survives, or opportunistic colonisers;

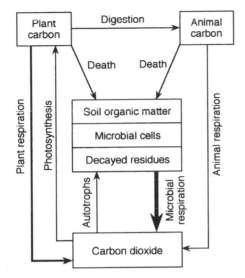

Figure 7.1. The role of the soil microbial community in the carbon cycle

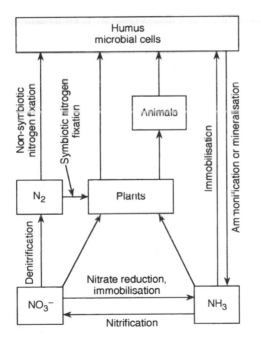

Figure 7.2. The role of the soil microbial community in the nitrogen cycle

- *Enrichment disturbance*, in which the carrying capacity of the soil is increased by the addition of nutrients from outside of the system.

There are numerous forms of human intervention which result in serious disruptions to the normal biological functioning of soils, requiring attention when new woodland planting is under consideration. Two examples of major disruption — the effects of soil storage and reinstatement, and pollution — are now considered in greater detail.

Soil storage and reinstatement

In the United Kingdom, opencast coal mining is principal among several industries involving widespread disruption of the soil–plant system. Careful management of remaining soil resources is essential in order to facilitate rapid restoration (see Moffat and Buckley, Chapter 6). There is now considerable evidence to the effect that opencast mining, and indeed any operation which moves soil in any way, has a major impact on the state and quality of the soil (Harris and Birch, 1992); and although many operations are designed with 'progressive' restoration, that is the soil is directly replaced onto another area, the vast majority still include a period of storage in large mounds. The effects of storage on soil biota are time dependent, and the processes may be divided into a number of phases:

- *Phase 1 (instantaneous)*. During the early period of soil storage a number of events occur simultaneously, arising from the immediate effects of soil structural disruption. A large proportion of the plant and fungal biomass is killed as a result of this process.
- *Phase 2 (1 to 3 months)*. During this phase there are major increases in the numbers of bacteria throughout the soil store and to a lesser extent a flush in the numbers of surviving fungi in the top part of the soil store. This is engendered by the availability of organic materials, killed by the construction process. At this time the onset of anaerobic conditions within the soil store is seen, in the deeper parts of the store, first as free oxygen, followed by nitrates and phosphates as each are utilised.
- *Phase 3 (6 months plus)*. By now three zones are well established within the soil store (Figure 7.3):
  - *The aerobic zone*, at the top of the store, where conditions begin to return to normal, particularly with the re-establishment of a vegetation cover. This soil is still, however, severely compacted in comparison to undisturbed controls.
  - *The anaerobic zone*, within the body of the store, characterised by very low numbers of micro-organisms (except anaerobes), seeds, mycorrhizal infectivity, high levels of ammonium and available metals.
  - *The transition zone* fluctuates between the two states, becoming anaerobic when filled with water, and aerobic when it dries out.

    The size of the various zones will be dependent on the textural class of the soil. Fine textured soils will have an anaerobic zone extending to as little as 20 cm from the surface of the store. Sandy soils in store may have no anaerobic zone at all until they reach several metres in height.

At the end of the storage period, which may last from as little as 3 months to as much as 20 years, the soils are seriously degraded. There is another 'bacterial peak' with a fresh supply of organic matter being killed during the store stripping and re-instatement

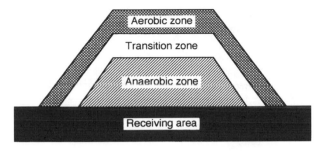

Figure 7.3. Development of zonation following soil storage

procedure. This leaves the soil biota in a very poor condition: low bacterial numbers, very poor fungal numbers and mycorrhizal infectivity undetectable, and low in available N. There is a great deal of work which needs to be done to improve the quality of the soil if it is to regain its former function, and not merely be a 'hydroponic' system, in constant need of chemical amendments and physical attention. This must include careful management of soil organic matter, and the encouragement of particular groups of organisms, by inoculation and introduction of appropriate primary producers.

Pollution

Another major impact on the quality of soils is that of pollution. This can be either in the form of on-site contamination, caused by the presence of contaminative industry, or by off-site sources, commonly atmospheric pollution. This type of pollution has a significant impact on soil chemistry, with the pollutants dominating reactions, leading to pH being altered well outside of the range in which plant growth is possible.

The biota is often profoundly affected by such contamination, leading to the disappearance of earthworms, micro-arthropods and other complex organisms. Populations of micro-organisms may often adapt, but this is usually at the expense of some aspect of their function; for example, in soils receiving sewage sludge containing heavy metals for years, the *Rhizobia* spp., responsible for fixing atmospheric nitrogen for clover, were still present but were incapable of fixing nitrogen, even when removed to unpolluted substrates.

SUCCESSIONAL PROCESSES ASSOCIATED WITH WOODLAND RESTORATION

Natural secondary succession

There are many systems that have undergone the process of secondary succession, simply as a result of being left to natural processes of invasion and establishment. Many former pasture and arable sites have been 'abandoned' and over the course of time woodlands have been re-established. In many important respects this is the best type of woodland restoration. Assemblages of plant species come to a quasi-steady state quite naturally, leading to the return of organisms essential for the soil functioning. When decomposer activity has been re-established, invertebrates are also able to invade and establish. This will lead to the stabilisation of soil structure and the formation of root and earthworm passages, forming drainage channels and the re-establishment of microbial communities (see Figure 7.4).

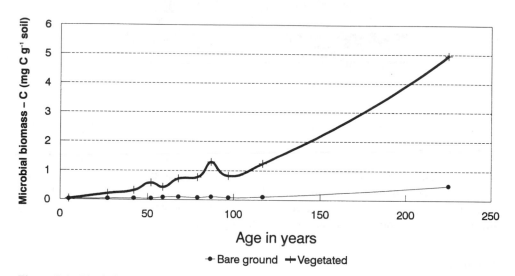

Figure 7.4. The influence of natural secondary succession on soil biotic processes: effect of vegetation on organic carbon accumulation on the Athabasca moraines. Redrawn from H. Insam and K. Haselwandter (1989), *Oecologia,* **79**, 174–178, and reproduced by permission from Springer-Verlag.

This approach, however, does have one serious drawback — time. If such programmes are to be fully supported then it is natural that people wish to see change in their own life-times. In many cases the basic soil conditions are not right and genetic stocks are lacking; it may be several hundred years before anything approaching an 'ancient woodland' is achieved. For this reason it is important that we intervene directly.

Active intervention

There are two types of site which need to be considered with respect to the state of the soils: those which are contaminated and those which are not. Contaminated soils require treatment which will be particular to the type of contaminant. The main approaches are: removal, containment and various forms of *in situ* treatment. With both removal and containment, the application of a new cover material (often 'clean fill') is necessary. Unfortunately it is usually devoid of organic matter, soil biota, and structure.

Where *in situ* treatment is carried out there are well-established physico-chemical means of removing inorganic and organic contaminants, such as thermal scrubbing and chemical oxidation. Unfortunately they also remove the functional organic components of the soil. In recent years there has been greater focus on using microbiological methods for dealing with organic contamination. The existing microflora, which has adapted to the contaminant, is enriched by addition of specific nutrients in order to enhance the decomposition process. This usually involves the addition of nitrogen and other nutrients but not carbon sources, as these will be used in preference to the contaminant which is typi-cally more difficult to degrade. Many successful clean-up operations have been achieved using this type of technology. It is not possible, however, to remove inorganic pollutants this way, and the best approaches remain the physico-chemical treatments.

In sites which have not been contaminated but merely disturbed, then the problems faced are restoration of nutrient cycling and structural stability. The degree of interven-tion will be dependent upon the degree of damage and the type of soil material available. The soils will be in various states of poor structure, nutrient deficiency and low genetic stocks. Following the total disturbance of mining, the rate of recovery of the newly estab-lished community will largely depend upon the quality of the restoration. The recovering habitat can suffer multiple stresses resulting from low levels of organic matter, including a deterioration of soil structure from crumb toward massive, and low rates of nutrient mineralization by a depleted and modified soil biota (worms are especially sensitive). Low levels of N, P, and other minerals are common, and the biota may be poisoned by residual pollutants or the products of spoil mineralisation and oxidation (e.g. Pb, Cu, or Al mobilisation or the production of acid from oxidising pyrite). Soil compaction, a common problem following soil replacement, will restrict root penetration and promote waterlogging, anaerobiosis and desiccation.

Restoring biotic function

Organic matter has to be added for three reasons: (1) to encourage the establishment of earthworms and other invertebrates, (2) to begin decomposer activity and (3) to stabilise soil structures. Without such additions restoration will be a very long-term process. Organic matter additions such as sewage sludge have been used to great effect, but it must be noted that they have traditionally brought with them the problems of heavy metal pollution. It may be that in the early days of restoration when organic matter is totally absent, then this is an acceptable 'trade-off'. There then needs to be a long phase where periodic additions of organic materials are made; this may take the form of a 'green manure' such as clover (*Trifoloium* spp.) or melilot (*Melilotus* spp.). It has been shown that by applying appropriate amendments it is possible to encourage the growth of tree species and perennial stocks during the new succession which is taking place (Figure 7.5).

Subsequent successional processes

When released from the constraints of human disturbance, sites will follow a successional path determined by the outcome of what Pickett and McDonnell (1989) termed a causal repertoire of successional forces. For the vegetation, the causal repertoire will include such things as further disturbances (drought, windthrow, disease, etc.), the composition of the pre-existing seed pool, and the impact of differences in species' dispersal efficiencies.

In the soil, a different repertoire operates. For example, hydrogen ion concentration is a successional force because of its enormous influence upon the solubility of many minerals, root growth and function, ectomycorrhizal formation and the composition and activity of the soil microflora and fauna. A change in pH during succession will itself depend upon the balance found between an array of mechanisms. Alkalising processes include the weathering of minerals that contain carbonates and hydroxides, the uptake of more anions (NO_3^-, SO_4^{2-}, $H_2PO_4^-$, Cl^-) than cations (K^+, Na^+, Mg^{2+}, Ca^{2+}) including the

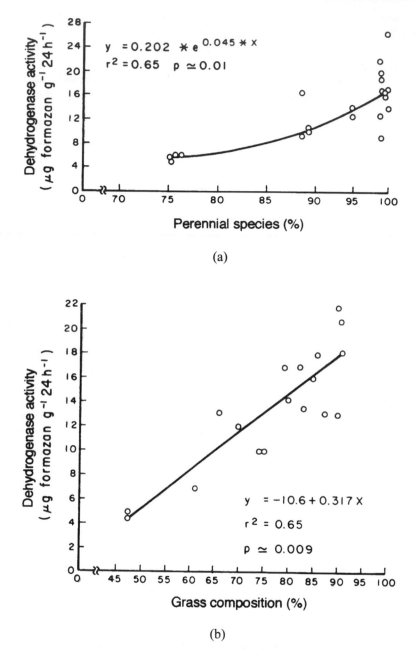

(a)

(b)

Figure 7.5. Relationship between plant species composition and soil dehydrogenase activity. (a) Relationship between dehydrogenase enzymatic activity and perennial species composition (grasses or lucerne, *Medicago sativa*) in an introduced seed mixture. (b) Relationship between dehydrogenase enzymatic activity and perennal grass composition in a native seed mixture. Reproduced by permission of Elsevier Science from M.E. Biondini *et al.* (1985), *Reclamation and Revegetation Research*, **3**, 323–342

uptake of NO_3^- in preference to NH_4^+ by plant roots (leading to excretion of HCO_3^- and OH^-), and the mineralisation of organic N to NH_4^+. Counteracting acidification is caused by the accumulation of organic matter or a change in its source (humus contains numerous weak and strong acidic functional groups), the formation of carbonic acid from CO_2 respired by roots and the soil biota, the mineralisation of organic matter to produce nitrate, sulphate and phosphate (oxidation of NH_4^+ to NO_2^- by *Nitrosomonas* spp. releases two H^+ ions), acid deposition in throughfall, fertiliser use, and leaching losses of Ca^{2+} and Mg^{2+} (Jenkinson, 1988; Rowell, 1988; Ineson, 1992; Marschner, 1992). The above inventory serves to illustrate that what we describe simply as hawthorn scrub or 30-year-old birch woodland is actually the integration of all ecosystem processes.

Describing succession as the outcome of a set of forces condenses an otherwise overwhelming array, and so clarifies matters in several ways. The rate of habitat creation can be maximised, and errors avoided, by focusing management and monitoring efforts on the most important factors controlling progress at a site. For example, the careful nurturing of a desired climax tree species on a restored mine or on derelict industrial land by planting, fertilisation, protection from competition and grazing by rabbits, and inoculation with mycorrizhae using soil transplants, may amount to nothing if after 15 years the trees' roots have penetrated no deeper than 15–20 cm into the compacted subsoil (Scullion, 1994) and the site is flattened by a winter storm; or, having been predisposed to infection by both waterlogging and droughting, most trees are killed by honey fungus, *Armillaria mellea*, that arrived along with the mycorrhizae in the carefully distributed sods. Disaster could have been avoided if drains had been installed, the profile was regularly fractured by subsoiling, and deep-burrowing worms re-established (Scullion, 1991). An alternative to the immediate replacement of trees may be to initially establish a dense community of competitive ruderals and stress-tolerant, tap-rooted ruderals to rapidly exploit any newly created fractures in the subsoil while injecting readily decomposable organic matter throughout the occupied profile (K. Willis, personal communication).

The habitat will continue to change while its repertoire of forces are imbalanced. With time, a new dynamic equilibrium state, the climax community, will form, but, like any perfect state, equilibrium will never fully be reached while fine scale disturbances such as tree falls continue to reset the successional clock, creating a mosaic of late-successional microhabitats. Succession was reversed over much of south-east England on the night of October 16, 1987 when a severe storm blew down 15% of the volume of all timber (Peterken, 1993).

The theory of successional forces uses a hierarchical structure of both population/community and process/functional attributes to model ecosystems. This approach avoids the perennial debate about whether succession is holistic (akin to the development of an organism, and so orderly, predictable, repeatable, increasingly in control of its environment, and ultimately stable), or whether it is individualistic (simply the product of a collection of individuals exhibiting a range of genetically determined strategies in an ever-changing environment, and characterised by randomness, unpredictability, and non-equilibrium). O'Neill *et al.* (1986) recognised that the population/community approach is partly a product of the historical development of ecology, whereby a concentration upon objects of human perceptual scale, such as higher plants and vertebrates, has led naturally to the basic subdivision of ecosystem structure into primary producers, consumers and decomposers. The population/community approach has particular problems dealing with the decomposer subsystem because essential soil components such as organic matter and nutrient cycling are normally considered to be outside the system (part of the environment). Also, the high level of redundancy that characterises most of the soil biota, and the dormant state of much of the microflora at any given time, upsets any attempt to predict community function from the species composition. Further, identification is difficult, not least because <10% of the biovolume of the microflora and only 0.1–1% of the visible bacteria will form colonies on nutrient agar (Olsen and Bakken, 1987). A process/functional approach to the soil also simplifies sampling because decomposition and nutrient cycling operate on a small scale. Care must be taken, however, not to try and divorce soil functioning from the vegetation that is supplying the flow of energy (as matter) and information, particularly when, as is common in woodlands, one species can dominate. O'Neill et al. (1986) proposed that it is quite feasible to marry an individualistic concept of the vegetation with a holistic concept of ecosystem function, and that a functional approach is particularly suited to the decomposer community. They recommended using a dual hierarchy of organism, population and community on the one hand, and functional component, ecosystem and biosphere on the other.

The problem of deciding whether a developing woodland has reached equilibrium can also be overcome by perceiving changes with time as having a hierarchical structure (O'Neill, 1989). Parts of the ecosystem that have similar rates of processing will belong in the same temporal level, while faster (lower level) parts will appear as noise that can be eliminated by averaging, and slower (higher level) parts will act as constants or boundaries. For example, a young tree in an emerging wood will, during several years of growth, perceive the pulses of N supplied by animal droppings as noise, while the amount of N released each year through mineralisation of organic matter would appear as a constant even though it would be gradually declining. Consequently, even though a woodland community will continue to change we can decide if equilibrium has been reached by selecting an appropriate time scale and an acceptable level of variation. A hierarchical approach to time helps to resolve the debate over whether the vegetation controls the biomass and activity of the decomposer community or whether plant growth is limited by the rate of nutrient cycling by the decomposers. Over the short term, from weeks to, perhaps, one or two years, plant growth will be limited by the nutrients supplied by the decomposer community. Decomposition may supply 70–85% of the annual growth requirement of N, P, K, Ca and Mg in temperate deciduous forest (Schlesinger, 1991). Over successional time scales, however, the soil community is controlled by the vegetation, through the quantity and nutritional quality of the organic matter supplied, the rate of symbiotic N fixation, storage of nutrients in long-lived plant parts, and other processes. Interestingly, the pendulum of dominance may finally swing back to the microflora due to the increasing reliance of ageing ecosystems upon the decomposition of organically bound soil P (Peet, 1992) and the central role of fungi and bacteria in the biogeochemical degradation of primary minerals (Walton, 1993).

Table 7.2 combines a functional/process approach with a hierarchical structure to generate a model of the decomposition subsystem tailored to changes in the soil during secondary succession. Characteristics that remain essentially unchanged during the transformation of grassland into woodland (spanning decades to a few centuries), such as primary mineral composition, topography and annual rainfall, have been omitted. Nested within a primary division of biotic and abiotic components are seven major groups: soil physical characteristics, soil chemical properties, nutrient and energy supply, decomposers, saprovores, predators/omnivores and root pathogens/grazers, each of which contains further subdivisions. The functioning of each level of the hierarchy can be understood by examining the next lower level. Exhaustive branching to reveal a complete structure of all species is unnecessary because the redundancy that characterises many lower level organisms will ensure continued functioning with a variety of species assemblages. Similarly, for abiotic components, the same rate of N mobilisation, for example, can be achieved through various configurations of mineralisation, deposition, immobilisation, denitrification, leaching, etc.

Table 7.2. Hierarchical structure of the soil decomposition subsystem, organised according to its process/functional attributes (functional measures of the biota could include biomass, activity, nutritional status or N mineralisation. C and N have been combined within the same section as both are derived from the atmosphere).

Physico-chemical			
Soil physical characteristics	Profile (O, A, B horizons)	Depth Texture: sand, silt, clay	
	Density	Bulk density: packing density, air capacity Mechanical impedance Water holding capacity	
	Aggregation	Crumb size distribution Crumb strength	
Soil chemistry	Ion exchange	Cation exchange capacity Anion exchange capacity Bonding strength	
	pH (or exchangeable H^+)	Acidifying processes	Accumulation of organic matter or change in source
			Mineralisation of organic matter
			Formation of carbonic acid from respired CO_2
			Acid deposition
			Use of ammonium fertiliser
			Leaching losses of Ca^{2+} and Mg^{2+}
		Alkalising processes	Weathering release of carbonates and hydroxides
			Uptake of more anions than cations by plant roots
			continued overleaf

Table 7.2. (*continued*)

Physico-chemical		
		Alkalising processes	Mineralisation of organic N to NH_4^+
	Redox potential	
	Toxicity	Development or abatement of toxicity from metal ions, H^+ or OH^-, organic pollutants or allelopathic chemicals
Nutrient and energy supply	C, N	Totals
		Degradability of organic matter: microbial biomass C/organic C, C/N, lignin/N, or decomposable plant material/resistant plant material
		Rate of N release: resultant of organic matter mineralisation (including nitrification), atmospheric deposition, fertilisation, root exudation — immobilisation, occlusion in stable humus, denitrification, leaching, NH_3 diffusion
	P, S	Totals
		Rate of release of available forms
	Bases (Ca, Mg, K)	Totals
		Exchangeable bases
	Micronutrients (Fe, Zn, Mn, Cu, Mo, B, Cl, Na)	Concentration if limiting
	Seasonal variables (monthly)	Available water (including months waterlogged)
		Temperature
Biota (food groups in brackets)		
Decomposers	Fungi	
	Bacteria including Actinomycetes	
Saprovores	Small (<2 mm)	Protozoa: amoebae, flagellates, ciliates (bacteria)
		Nematodes (bacteria, fungi)

continued

Table 7.2. (*continued*)

	Biota (food groups in brackets)	
	Small (<2 mm)	Microarthropods: mites, spring-tails (fungi, detritus and bacteria)
	Medium to large (>2 mm)	Worms: enchytraeids (fungi, bacteria, detritus), lumbricids (detritus)
		Arthropods: insects, millipedes and woodlice (detritus)
		Molluscs: slugs, snails (plant residues, detritus)
Predators/omnivores	Small (<2 mm)	Nematodes (protozoa, nema-todes, bacteria)
		Microarthropods: mites (proto-zoa, nematodes, springtails, mites), springtails (nematodes)
	Medium to large (>2 mm)	Arthropods: insects, spiders, centipedes, etc.
		Vertebrates: moles
Root pathogens/grazers	Pathogens	Fungi
		Bacteria
		Nematodes
	Grazers	Arthropods: insect larvae

CHANGES IN THE DECOMPOSITION SUBSYSTEM DURING WOODLAND SUCCESSION

Succession at Hainault and Epping Forests

At Hainault and Epping Forests, Essex, UK, the transformation of the decomposition subsystem during succession was monitored using subdivisions of two of the three abiotic groups (pH, and C and N dynamics) and one of the four biotic groups (the decomposer community). At each location, a toposequence of three communities growing on London Clay was sampled in winter, by taking 15 replicate soil cores, to a depth of 30 cm, from each community. Each toposequence comprised:

● A long established meadow site
● A 30- to 31-year-old birch/oak (*Betula pendula./Quercus robur*) woodland site growing on abandoned meadow or pasture
● An ancient woodland site of pollarded hornbeams (*Carpinus betulus*) and standard oaks (*Quercus robur*).

For the two abiotic subdivisions, pH was measured in a soil and water slurry, while C and N dynamics were assessed using total soil N, soil organic C, C/N ratio, and the microbial biomass C/organic C ratio.

Three measures were made of the decomposer community. Total microbial biomass was estimated from the adenosine triphosphate (ATP) concentration in the soil. ATP is a fundamental energy source for cellular reactions and can be used as a measure of living biomass because it occurs in roughly constant proportions in most organisms. The weight of microbial biomass C is obtained by multiplying ATP concentration by 168.

Fungal biomass was estimated from the concentration of ergosterol, a fungal sterol. Due to the variability in ergosterol production in different fungal species, no attempt was made to convert its concentration into an estimate of fungal biomass. Instead, the relative contribution of the fungi to the total microbial biomass was gauged from the ergosterol/ATP ratio.

Microbial activity was measured using dehydrogenase enzyme activity. This measures the activity of a collection of endogenous enzymes that catalyse the breakdown of organic molecules. Activity is measured indirectly from the rate of formation of a red dye, tri-phenyl formazan (TPF), over 24 h. Dehydrogenase activity has been correlated with the rate of O_2 uptake and CO_2 release from the soil, and so can be used as an index of microbial metabolic activity.

Grassland processes

The extensive root systems associated with grassland communities exude large amounts of soluble C in various forms (sugars, organic acids, amino acids, enzymes, metal ion chelators and signal molecules) to feed symbiotic mycorrhizae and to 'prime' bacteria and fungi in the rhizosphere to degrade resistant organic matter, so releasing N, P, S, and other minerals. A ubiquitous need for N is revealed, even in rich, grazed floodmeadows, by the dark green tussocks that flag points of cow defecation. Readily decomposable plant material is also supplied by the sloughing of root cap cells, autolysis of epidermal and cortical cells, production of mucigel, and the rapid turnover of root hairs and fine roots. Further, the seasonal death of the annuals and many roots of the perennials provides a winter pulse of substrate to the microbes and their attendant population of saprovores and predators. Earthworms flourish and their activity greatly modifies the soil: burrowing and mixing of the profile assists water movement and aeration, and transports microbes; incorporation of surface litter into the soil accelerates its decomposition and prevents the accumulation of a litter layer, retarding the development of acidity; and their mutualistic relationship with ingested microbes 'primes' organic matter decomposition. A large earthworm population may attract moles (*Talpa europaea*), further enhancing soil mixing, aeration and water infiltration.

The large biomass and activity of grassland microbial communities is revealed in Figure 7.6. Average microbial biomass of the top 30 cm of the soil in meadows at Hainault and Epping was 0.75 mg biomass C ml^{-1} soil, 3.6 times the ancient woodland value, and equal in fresh weight of biomass to an astonishing 450 sheep ha^{-1} (using the conversion of Brookes *et al.*, 1985). Average microbial activity, at 535 µg TPF 24 h^{-1} ml^{-1} soil, was almost five times higher than under ancient woodland. The plentiful supply of root exudates and readily decomposable plant material is reflected indirectly by the higher level of total soil N (average meadow value of 2.9 mg ml^{-1} soil) and lower soil C/N ratio (average of 11.6), and directly by the much higher carrying capacity of the organic matter; C within the living microbial biomass accounted for 2.25% of the total organic C content of the soil, a level 4.0 times higher than under ancient woodland. Grassland fungal biomass, with hyphal lengths in the order of 0.5–2 km g^{-1} dry soil, normally comprises 50–75% of the total microbial biomass, and becomes even more dominant in later successional stages. The relative fungal biomass, estimated from the ratio of ergosterol to ATP, appears to increase dramatically with woodland development (Figure 7.6). However, the magnitude of this increase may have been influenced by changes in the ergosterol concentration in fungal tissues during succession. Soil pH (5.4 and 6.0 at Hainault and Epping meadows, respectively) was slightly acidic.

Woodland colonisation

Cessation of cutting and grazing allows scrub and woodland species to invade. Domination by shrubs and trees drives a major transformation of the soil profile as litter fall and feeder root proliferation at the soil surface increasingly replaces the *in situ* injection of organic matter throughout the profile by grassland species. Without constant investment, the organic matter capital accumulated at depth declines. Easily degradable forms such as plant residues and microflora will decompose rapidly, with half-lives of months to a few years, while the high molecular weight compounds and various humus fractions formed from their decomposition are more stable, with turnover times ranging from decades to several thousand years (Jenkinson, 1988; Parton *et al.*, 1988). Humus formed by mixing barley and maize straw into loam decayed with a half-life of 5–7 years after an initial 3-month period of rapid decay (Sorensen, 1983). Averaging across all forms, Jenkinson (1988) assigned organic C a half-life of 25 years beneath grassland. Consequently, after 20 years of woodland succession, much of the initial legacy of readily decomposable organic matter at depth, and the microbial and faunal community that it sustained, would have disappeared. Remnant humus will function more as a site of cation exchange than as a reservoir of energy and nutrients, and the decline in microbial and earthworm activity will lessen the rate of soil crumb formation, creating a more massive soil structure.

Accumulation of litter

Accumulation of surface litter attracts mycorrhizal feeder roots to the surface O_h (humic organic) and A_h (humic mineral) horizons. The initially high C/N, C/P and C/S ratios of the litter result in nutrient immobilisation during the early stages of decay, but as decomposition progresses the C content is reduced and N, P and S are mineralised. Below the surface litter and fragmentation zones, a 5–10 cm deep humus layer will develop that can be so ramified with mycorrhizal roots that it may be peeled off like a deep carpet. While falling litter controls the location of the feeder root network, a significant proportion of each year's organic matter will be supplied by the rapid turnover of the roots themselves. In cold, temperate, broadleaf woodland, inputs from fine roots amounted to 2280 kg ha^{-1} yr^{-1} while litterfall accounted for 3854 kg ha^{-1} yr^{-1}, and, in beech forest, fine root turnover supplied 29% of the total of above- and below-ground annual inputs (Vogt *et al.*, 1986).

Scrub litter will be composed almost entirely of leaves, but, as the tree canopy emerges, the proportion of woody inputs will increase, and long-term immobilisation of N becomes more common during the slow decay of tree trunks, larger branches and roots. Before abscission, the nutritional quality of leaves is reduced by the reabsorption of significant amounts of nutrients. In temperate hardwood forest, 30% of N and P, and 2–4% of K and Mg are withdrawn from leaves during senescence (Schlesinger, 1991). This internal cycling of N and P will increase during woodland development and is likely to be stronger in trees adapted to nutrient poor soils. As the woodland develops, and woody structural tissue accumulates, the C/N and C/P ratios of the total plant biomass

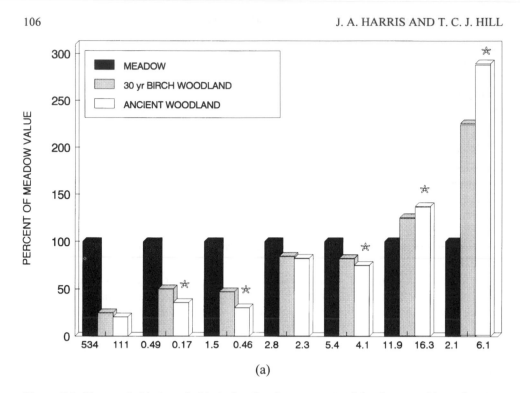

(a)

Figure 7.6. Changes in biotic and abiotic functional components of the decomposition subsystem in toposequences of succession from meadow to ancient pedunculate oak *(Quercus robur)*/hornbeam *(Carpinus betulus)* woodland at (a) Hainault Forest, Essex, and (b) Epping Forest, Essex. Absolute values for each measure are given for meadow and ancient woodland. Microbial activity was estimated from dehydrogenase enzyme activity (μg TPF 24 h^{-1} ml^{-1} soil), and microbial biomass carbon from the ATP concentration (μg ATP * 168.5 ml^{-1} soil). Units for organic C and total soil N are mg ml^{-1} soil. Relative fungal biomass is expressed as the ratio of ergosterol/ATP concentration. An asterisk indicates that birch *(Betula* spp.) oak *(Quercus* spp.) woodland and ancient woodland values were significantly different (*t*-test, $p < 0.05$)

will increase. Vitousek *et al.* (1988) reported an average value of 165 for C/N and 1384 for C/P in temperate, broadleaf, deciduous forests. At Hainault Forest, the C/N ratio of the top 5 cm of the soil (O$_h$ and A$_h$ horizons) was 19.0 in ancient woodland compared with 13.4 in adjacent meadow (A$_h$ only). The large standing biomass also acts as a repository of nutrients, especially Ca (a structural component of cell walls) and Mg, thereby permanently withdrawing them from the decomposition subsystem. Total N in the top 30 cm of the soil of ancient woodland was 82% and 68% of the meadow values at Hainault and Epping, respectively (Figure 7.6).

Degradation of litter

Scrub and woodland litter is also intrinsically more difficult to degrade due to the abundance of lignified and secondarily thickened cells in the tissue. In a model of C, N, P and S dynamics in grassland soils, Parton *et al.* (1988) assigned C a turnover time of 1.5 years if present as nonlignified, structural plant residue, while C within lignified structural tissue was given a turnover time of 25 years. Jenkinson *et al.* (1992) characterized

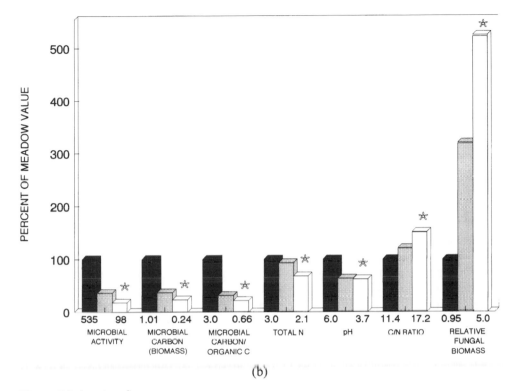

Figure 7.6. *(continued)*

organic inputs in grassland as having a ratio of decomposable to resistant plant material of 1.44, while the ratio in 100-year-old woodland was much lower, at 0.25. Under the same conditions, decomposable plant material would decay 33 times faster than resistant plant material. Other constituents, such as polyphenols and calcium oxalate crystals in cell vacuoles, epicuticular waxes, callose plugs in sieve tubes, and suberin-impregnated bark cells further inhibit decomposition. Harrison (1971) found that tannins leaching from decaying oak leaves inhibited the growth of 14 out of 19 fungi isolated from the litter. The progressive fall in the palatability of plant residues during woodland succession is reflected by a drop in the ratio of microbial biomass C/total organic C from 2.25% in meadow to 0.83% in 30-year-old birch woodland, and finally to 0.56% in ancient *Quercus-Carpinus* woodland (Figure 7.6).

Accumulation of humus

Dissociation of abundant acid (COOH) and phenolic (OH) groups in the accumulating humus produces a rain of hydrogen ions from the soil surface. Hydrogen ions and K^+ ions (leached from the canopy and mineralised from decaying litter) combine to elute less mobile Ca^{2+} and Mg^{2+} ions from surface horizons, which, together with their immobilisation within long-lived structural tissues, lowers the level of base saturation. A higher concentration of these monovalent cations also offsets the flocculation of colloids and so promotes the illuviation of clay and more mobile humus fractions, such as N-, P- and S-rich fulvic acids, to lower soil layers. At Epping Forest, the pH of the top 30 cm of the profile fell from a meadow value of 5.9 to 4.1 beneath *Betula-Quercus* woodland: a 75-fold increase in H^+ concentration in only 30 years. Acidic conditions further inhibit decomposition of the increasingly unpalatable substrate. In time, however, the rate of litter accumulation will level off as the slow decay of the deep organic layer matches the yearly C inputs. Organic C content of the soil was similar in all sites.

Successional changes in the soil biota

Not surprisingly, the collection of simultaneously acting successional forces outlined above rapidly lowers the biomass and activity of the microflora, and changes its composition (Figure 7.6). At Hainault Forest, microbial biomass and activity beneath 14-year-old scrub was 38% and 47%, respectively, of the meadow levels. Fungi dominate woodland microbial biomass due to a combination of their tolerance of acidic conditions (Alexander, 1977; Harris, 1988) and their ability to degrade substrates with high C/N and lignin/N ratios; slow growing basidiomycetes prevail as lignin degraders. Grime (1988) proposed that the primary strategy of the vegetation will force a similar strategy upon the decomposer community, so that with a shift from competitive perennials to stress-tolerant, competitive trees the 'stress tolerance among primary producers begets stress tolerance in herbivores and decomposers'. Fungi span a wide range of strategies from ruderal, competitive to stress tolerating, so enabling them to dominate both the grassland and woodland decomposer communities.

Most earthworms avoid low pH and litter with high C/N and lignin/N ratios (Swift *et al.*, 1979). Earthworms (*Lumbricus terrestris*) were absent from three out of four woodland sites at Epping Forest, while moles (*Talpa europaea*) which feed primarily upon them were not found in mature woodland (Wheeler and May, 1992). In formerly arable, 300-year-old woodland in central Lincolnshire, Peterken (1993) observed that the lack of strong acidity enabled *L. terrestris* to persist, and that their continued incorporation of surface litter into the soil has prevented the accumulation of a litter layer.

In Figure 7.6 it is evident that after 30 years of succession the decomposition subsystem beneath *Betula-Quercus* woodland was broadly similar to that of ancient woodland. Notably, however, in all variables measured at Epping Forest and in all but two at Hainault Forest, significant differences were found between *Betula-Quercus* woodland and ancient *Quercus-Carpinus* woodland. Cluster analysis can be used to combine the discriminating power of all seven variables (Figure 7.7), and quantify the separation between ancient woodland and a range of early and late successional communities.

MEASURING SUCCESS

One of the principal questions facing the restorationist is 'how will I know when I have succeeded?'. This has traditionally relied on the presence of community assemblages similar to the target sites. There can then be adjustment by plantings, selective cutting, and measures designed to encourage the invasion of specific fauna. Unfortunately many of these procedures depend on the expertise of the investigator and are extremely time consuming. This type of analysis cannot indicate whether the system is truly functioning as it should.

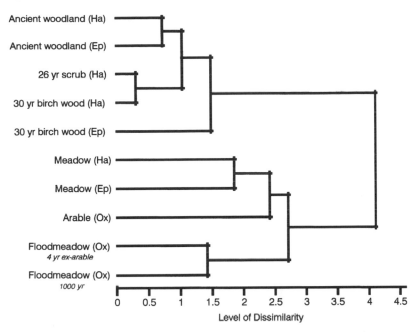

Figure 7.7. Hierarchical cluster analysis of meadow, scrub and woodland communities at Hainault and Epping Forests, Essex, and arable, restored floodmeadow and 1000-year-old floodmeadow at Oxford, using the same seven measures of the soil decomposition subsystem presented in Figure 7.6. Analysis was performed on a matrix of standardised Euclidean distances, with clusters generated using unweighted pair grouping

Recent work has indicated that there is a means by which an objective assessment of the success or otherwise of restoration may be made (Harris and Birch, 1992). It is clear that there are profound and consistent changes in the soil microbial community as a result of the types of stress and disturbance outlined. As a result of this it has been shown that by using measurement of the status of the soil microbial community a clear indication of the status of the system can be given.

Microbiological indices

Microbiological indices are proving to be extremely effective in indicating the extent to which soils have been damaged, and what effect management programmes have on repairing ecosystem damage. Bentham *et al.* (1992) have demonstrated that by using just three characteristics of the soil microbial community, a discrimination between functional sites may be achieved which is far superior to conventional measurements. These characteristics are:

- *Size:* the amount of biomass in the system, which may be determined directly in the form of carbon, or indirectly by extraction of ubiquitous cellular components such as adenosine triphosphate (ATP).
- *Activity:* a measure of the rate of turnover of materials within the system, and export/import of nutrients.
- *Composition:* the degree of biodiversity within the system, which may be related to

the relative proportions of types of metabolic groups, or broad divisions such as the percentage of the total biomass which is fungal.

These characteristics have allowed the progress of restoration programmes to be followed very closely, as well as indicating points of stability (Figure 7.8).

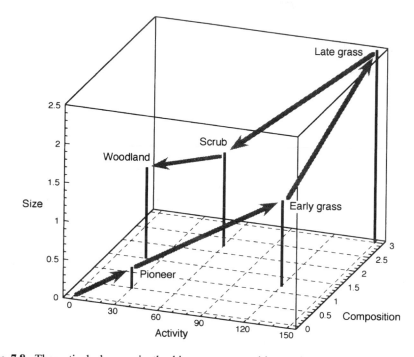

Figure 7.8. Theoretical changes in the biomass, composition and activity of the soil microbial community during succession to woodland. Size = microbial biomass; activity = microbial respiration; composition = fungal biomass

From an approach using microbiological indices, a thermodynamic model of these systems could be developed which would incorporate elements of the National Vegetation Classification model (Rodwell, 1991), i.e. incorporation of the three metabolic groups (primary producers, consumers and decomposers, plus detritivores if appropriate). This would allow prediction of not only effects of environmental perturbation on systems, but would also provide for a prescription for treatment, in terms of inputs of nutrients, organic matter, cultivation energy and genetic stocks, to create woodland communities.

Future prospects and research

A key factor in the re-establishment of woodlands is the need to integrate models of succession and stability at all levels, i.e. above- and below-ground structure and function, in order to develop an objective measurement of 'ecological soundness'. This in turn needs to be tied to practical examples of good practice and success. More attention needs to be paid to the importance of the soil in woodland re-establishment, and a research programme focused on methodologies designed specifically to enhance soil function.

REFERENCES

Alexander, M. (1977). *Introduction to Soil Microbiology*. Wiley, New York.

Bentham, H., Harris, J.A., Birch, P. and Short, K.C. (1992). Habitat classification and soil restoration assessment using analysis of soil microbiological and physico-chemical characteristics. *Journal of Applied Ecology*, **29**, 711–718.

Biondini, M.E., Bonham, C.D. and Redente, E.F. (1985). Relationships between induced successional patterns and soil biological activity of reclaimed areas. *Reclamation and Revegetation Research*, **3**, 323–342.

Brookes, P.C., Powlson, D.S. and Jenkinson, D.S. (1985). The microbial biomass in soil. In: Fitter, A.H., Atkinson, D., Read, D.J. and Usher, M.B. (eds). *Ecological Interactions in Soil: Plants Microbes and Animals*. Special Publication No. 4 of the British Ecological Society. Blackwell Scientific Publications, Oxford, 123–125.

Grime, J.P. (1988). Fungal strategies in ecological perspective. *Proceedings of the Royal Society of Edinburgh B*, **94**, 167–9.

Harris, J.A. and Birch, P. (1992). Land reclamation and restoration. In: Fry, J.C., Gadd, G.M., Herbert, R.A., Jones, C.W. and Watson-Craik, I. (eds). *Microbial Control of Pollution*. Society for General Microbiology, Symposium 48, Cambridge University Press, Cambridge, 269–291.

Harris, P.J. (1988). The microbial population of the soil. In: Wild, A. (ed.). *Russell's Soil Conditions and Plant Growth*. Longman Scientific and Technical, Harlow, 449–471.

Harrison, A.F. (1971). The inhibitory effect of oak leaf litter tannins on the growth of fungi, in relation to litter decomposition. *Soil Biology and Biochemistry*, **3**, 167–2.

Ineson, P. (1992). Forest soil biology: impossible challenge or open market? In: Teller, A., Mathy, P. and Jeffries, J.N.R. (eds). *Responses of Forest Ecosystems to Environmental Changes*. Elsevier Applied Science, London, 165–175.

Insam, H., and Haselwandter, K. (1989). Metabolic quotient of the soil microflora in relation to plant succession. *Oecologia*, **79**, 174–178.

Jenkinson, D.S. (1988). Soil organic matter and its dynamics. In: Wild, A. (ed.). *Russell's Soil Conditions and Plant Growth*. Longman Scientific and Technical, Harlow, 564–607.

Jenkinson, D.S., Harkness, D.D., Vance, E.D., Adams, D.E. and Harrison, A.F. (1992). Calculating net primary production and annual input of organic matter to soil from the amount and radiocarbon content of soil organic matter. *Soil Biology and Biochemistry*, **24**, 295–308.

Marschner, H. (1992). Nutrient dynamics at the soil-root interface (rhizosphere). In: Read, D.J., Lewis, D.H., Fitter, A.H. and Alexander, I.J. (eds). *Mycorrhizas in Ecosystems*. CAB International, Wallingford, 3–12.

O'Neill, R. V. (1989). Perspectives in hierarchy and scale. In: Roughgarden, J., May, R.M. and Levin, S.A. (eds). *Perspectives in Ecological Theory*. Princeton University Press, Princeton, 140–157.

O'Neill, R.V., DeAngelis, D.L., Waide, J.B. and Allen, T.F.H. (1986). *A Hierarchical Concept of Ecosystems*. Monographs in Population Biology 23. Princeton University Press, Princeton.

Olsen, R.A. and Bakken, L.R. (1987). Viability of soil bacteria: optimization of plate-counting technique and a comparison between total counts and plate counts within different size groups. *Microbial Ecology*, **13**, 59–74.

Parton, W.J., Stewart, J.W.B. and Cole, C.V. (1988). Dynamics of C, N, P and S in grassland soils: a model. *Biogeochemistry*, **5**, 109–131.

Peet, R.K. (1992). Community structure and ecosystem function. In: Glenn-Lewin, D.C., Peet, R.K. and Veblen, T.T. (eds). *Plant Succession: Theory and Prediction*. Chapman and Hall, London, 103–151.

Peterken, G.F. (1993). *Woodland Conservation and Management*. Chapman and Hall, London.

Pickett, S.T.A. and McDonnell, M.J. (1989). Changing perspectives in community dynamics: a theory of successional forces. *Trends in Ecology and Evolution*, **4**, 241–245.

Rodwell, J.S. (ed.) (1991). *British Plant Communities*, vol. 1: *Woodlands and Scrub*. Cambridge University Press, Cambridge.

Rowell, D.L. (1988). Soil acidity and alkalinity. In: Wild, A. (ed.). *Russell's Soil Conditions and Plant Growth*, Longman Scientific and Technical, Harlow, 844–898.

Schlesinger, W.H. (1991). *Biogeochemistry: an Analysis of Global Change*. Academic Press, San Diego.

Scullion, J. (1991). Re-establishing earthworm populations on former opencast mining land. In: Davies, M.C.R. (ed.). *Land Reclamation: an End to Dereliction?* Elsevier Applied Sciences, London, 377–386.

Scullion, J. (1994). Land Management. In: Scullion, J. (ed.). *Restoring Farmland After Coal.* British Coal Opencast, Mansfield, 36–46.

Sorensen, L.H. (1983). Size and persistence of the microbial biomass formed during the humification of glucose, hemicellulose, cellulose and straw in soils containing different amounts of clay. *Plant and Soil*, **75**, 121–130.

Swift, M.J., Heal, O.W. and Anderson, J.M. (1979). *Decomposition in Terrestrial Ecosystems.* Blackwell Scientific Publications, Oxford.

Vitousek, P.M., Fahey, T., Johnson, D.W. and Swift, M.J. (1988). Element interactions in forest ecosystems: succession, allometry and input-output budgets. *Biogeochemistry*, **5**, 7–34.

Vogt, K.A., Grier, C.C. and Vogt, D.J. (1986). Production, turnover, and nutrient dynamics of above- and below-ground detritus of world forests. *Advances in Ecological Research*, **15**, 303–377.

Walton, D.W.H. (1993). The effects of cryptograms on mineral substrates. In: Miles, J. and Walton, D.W.H. (eds). *Primary Succession on Land.* Blackwell Scientific Publications, Oxford, 33–53.

Wheeler, A. and May, A. (1992). The status and distribution of the mole in Epping Forest. *The London Naturalist*, **71**, 151–156.

8 Creating Woodlands: to Plant Trees or Not ?

R. HARMER and G. KERR

INTRODUCTION

Establishment of woodland represents a marked and probably long-term change to the ecology of any site and should not be considered lightly. However, once the decision has been taken a number of methods of achieving new woodland are available, of which planting is only one. The form of the woodland that is to be created will be important, but a woodland can be defined in many ways, for example, 'a wooded area' (Ford-Robertson, 1971) or 'land covered with trees with clear spaces between them (i.e. not as dense as a forest)' (Collin, 1992). In this review a woodland is defined as an area whose vegetation is dominated by trees and other woody plants with no more than 10% open space.

There are many reasons for wishing to establish woodlands but these may be ill-defined, perhaps no more than a wish to create woodland because it is perceived as a 'good thing to do' or due to the availability of grant aid. However, it is helpful at an early stage to attempt to clarify objectives and priorities, despite possible inherent difficulties:

- There may be several, perhaps conflicting, objectives, e.g. to provide for amenity and to maximise economic return.
- The long-term use of the land may not be certain.
- Flexibility must be retained to allow changes in objectives.

Despite such difficulties the attempt to clarify objectives and decide priorities is the single most helpful step to resolve questions on the form of woodland to be created and the most appropriate way to achieve this goal. The method of woodland creation must be appropriate to the resources available and suitable for the site on which the woodland is to be created. In general the realisation of complex and diverse objectives may be expensive to achieve and will often require a high level of professional management. Hence, it is probably unrealistic for a farmer, who has little or no experience of forestry, to attempt the creation of new native woodlands using the complex planting patterns proposed by Rodwell and Patterson (1994). Likewise it would be impractical to attempt to grow veneer quality oak on poor quality sites such as reclaimed colliery spoil or derelict land as these usually have a substrate with poor physical, water and nutrient properties (see Moffat and Buckley, Chapter 6) and the establishment of tree cover, of any description, can be difficult.

The Ecology of Woodland Creation. Edited by Richard Ferris-Kaan.

There are two broad options for woodland creation which will be considered: establishment of woodlands from seed and establishment of woodlands by planting.

ESTABLISHMENT OF WOODLANDS FROM SEED

Natural regeneration

Natural regeneration, which is the process of restocking woodland from seeds produced and germinated *in situ*, is well known in theory, if not practice, by most foresters. It forms an integral part of many silvicultural systems and has often been described in silvicultural texts (Matthews, 1989; Schlich, 1904; Smith, 1986) and recently in more practical guides (Evans, 1988; Hibberd, 1991; Kerr and Evans, 1993). As silvicultural systems based on natural regeneration usually cause less obvious environmental disturbance than clearfelling systems, they are thought to be more ecologically sound, but successful use of natural regeneration often involves a rigorous silvicultural regime including procedures such as ploughing, scarifying and control of the understorey and ground flora. Well-known examples of natural regeneration include the selection system developed for beech and silver fir under alpine conditions, and the shelterwood system for oak in northern France (Benezit, 1991; Matthews, 1989; Teissier du Cros, 1987). With the possible exception of the Chiltern beechwoods, natural regeneration has never been an important method of restocking in Britain, and there are at present relatively few places where large areas of forest are managed with the intention of using this method of restocking. Notable exceptions are the Ebworth estate in Gloucestershire, England, which provides a good example of a group selection system for beech and the Forest of Dean, Gloucestershire, England, which is successfully using a shelterwood system for oak (Anon, 1990).

Recently a combination of catastrophic windblow and changes in forestry grant schemes in Britain has encouraged the use of natural regeneration; a survey in 1993 showed that on 75% of sites, use of natural regeneration was opportunistic rather than carefully planned (Harmer and Kerr, unpublished) and many sites had large numbers of small seedlings indicating that the potential for successful restocking by natural regeneration is high. However, unless these sites are carefully managed, excessive browsing and luxuriant weed growth may cause high seedling mortality.

Sites undergoing regeneration usually have some remaining parent trees or are very close to another source of seeds. A study of 30 regenerating sites in southern England has shown that the presence of parent trees does not always lead to natural regeneration (Table 8.1). A total of 23 species of parent tree were present but only 16 of these were found as seedlings on the same sites. Some parents were found on many sites whereas others were rare. Of the parents found on more than 10 sites, birch (*Betula* spp.) was the only tree which was always present as a seedling on the same sites; parents of *F. sylvatica* and sweet chestnut (*Castanea sativa*) were often found without seedlings. The reasons for the absence of seedlings may include:

● Failure to produce enough seed: important for masting species such as *F. sylvatica* and *Quercus* spp.
● Seed predation: also important for beech and oak.
● Competition from other vegetation for water, light and nutrients: important for *Betula* spp., ash (*Fraxinus excelsior*), sycamore (*Acer pseudoplatanus*).
● Browsing damage: affects most species.

Table 8.1. Forestry Authority survey of natural regeneration in southern England: species of parent trees present at 30 sites and percentages of sites with seedlings

	Sites with parent		Sites without parent	
	Number	% with seedlings	Number	% with seedlings
Oak (*Quercus* spp.)	26	88	4	25
Ash (*Fraxinus excelsior*)	24	71	6	0
Birch (*Betula pendula*)	18	100	12	17
Beech (*Fagus sylvatica*)	20	50	10	10
Holly (*Ilex aquifolium*)	15	73	15	27
Hawthorn (*Crataegus monogyna*)	14	64	16	19
Sycamore (*Acer pseudoplatanus*)	17	88	13	15
Sweet chestnut (*Castanea sativa*)	12	42	18	0
Field maple (*Acer campestre*)	11	55	19	11
Willow (*Salix* spp.)	9	22	21	24
Rowan (*Sorbus aucuparia*)	8	75	22	5
Wild cherry (*Prunus avium*)	9	11	21	0
Elm (*Ulmus* spp.)	5	0	25	0
Alder (*Alnus glutinosa*)	3	0	27	0
Aspen (*Populus tremula*)	2	50	28	0
Hornbeam (*Carpinus betulus*)	3	100	27	4
Crab apple (*Malus sylvestris*)	1	0	29	0
Horse chestnut (*Aesculus hippocastanum*)	1	0	29	0
Lime (*Tilia* spp.)	1	0	29	0
Poplar (*Populus* spp.)	1	0	29	0
Whitebeam (*Sorbus aria*)	2	50	28	4
Wild service tree (*Sorbus torminalis*)	1	0	29	0
Norway maple (*Acer platanoides*)	1	100	29	0

Successful use of natural regeneration will often require consideration of some or all of these factors. Problems are not always associated with securing the initial crop; on some sites regeneration is prolific and the density of stems present can make management difficult and expensive. In general, natural regeneration is most readily achieved on sites with poor, infertile, well-drained soils where competing vegetation is sparse. In contrast, natural regeneration is often more difficult on moisture retentive, heavy, fertile soils where growth of competitive vegetation is luxuriant (see Moffat and Buckley, Chapter 6). This is demonstrated in the Forest of Dean, where oak regeneration is successfully achieved on the poor, acid soils which occur over most of the forest, but is disappointing on the smaller area of fertile soil, despite the abundance of seed.

Natural colonisation

Natural colonisation is the process of woodland establishment on previously unwooded sites by natural dispersal of seeds. This method, which was recognised as having potential for woodland creation in the middle of the 18th century (Ellis, 1745), is probably the closest to a natural succession that can be achieved. However, in many cases it is unlikely to create established woodland in a timespan as short as that achieved by planting.

Ellis (1745) suggested that new oakwoods could be created in ploughed fields and meadows near to existing woods by bird-dispersed acorns and windblown ash keys, sallow (*Salix caprea*) and aspen (*Populus tremula*). Similar observations were reported by Mellanby (1968), who found that reasonable numbers of *Quercus* seedlings could establish within 1–2 years of abandoning cultivated ground that was near to woodland. The successional changes following abandonment of agricultural land have been studied frequently in North America (Gill and Marks, 1991; Hanks, 1971; Pickett, 1982; Raup, 1940) and results suggest that while woodland may eventually form, the time scales involved are often very long. Study of an old-field in Michigan showed that despite being only 300-400 m from a stand of *Quercus* spp., successful establishment of oak only began 15 years after abandonment (Harrison and Werner, 1984). Colonisation was slow on an old-field in Ontario which was 5.4 ha in area and surrounded on three sides by woodland. Fifty years after abandonment there were 12 tree species present, but only four were established as large trees, the remainder being seedlings or saplings: overall tree cover was very low, but two large groves of trembling aspen (*P. tremuloides* Mich.) with a total area of about 650 m^2 had developed (Maycock and Guzikowa, 1984). Timescales in Britain are similar, for example the scrub which developed on the South Downs following abandonment after 1870 was not invaded by trees until about 50 years later (Watt, 1924). There are few detailed, published records but those that exist illustrate the spatial and temporal differences which can be found during recolonisation.

Perhaps the best known British examples of long-term colonisation of agricultural land are the Broadbalk and Geescroft wildernesses at the Rothamsted experimental station. In 1882 an area of *c.* 0.2 ha in the Broadbalk field was abandoned after 39 years under continuous wheat; the area was bounded by a hedge on the west and the Permanent Wheat Experiment on the east. Early observations of vegetation changes concentrated on the herbaceous component but by 1894 young oak, ash, hazel (*Corylus avellana*), hawthorn (*Crataegus monogyna*) and rose (*Rosa* spp.) were growing near the hedge (Lawes, 1895). The trees and shrubs were grubbed from half of the area in about 1900 when trees began to dominate the lower growing plants (Witts, 1964; Jenkinson, 1970). A survey in 1913–14, 30 years after abandonment, showed that the wooded areas had become a dense thicket of trees and shrubs with bramble (*Rubus fruticosus*), which was similar in type to 'damp oak woods' or 'oak-hazel woods' (Brenchley and Adam, 1915). In 1945 a survey of the wilderness trees gave the first information on the relative abundance of the different species present (Witts, 1964): abundance declined in the order *Crataegus monogyna* > *Fraxinus excelsior* > *Quercus robur* > field maple (*Acer campestre*) > holly (*Ilex aquifolium*) > *Acer pseudoplatanus* > cherry (*Prunus* spp.) > *Corylus avellana* > blackthorn (*Prunus spinosa*) > elder (*Sambucus nigra*); there were also about 40 *A. pseudoplatanus* saplings (unless authorities are given nomenclature follows Stace, 1991). A further survey made in 1960 showed that these saplings had grown and that many *Corylus avellana* and *Crataegus monogyna* were dying, which was probably due to the dense shade cast by the taller trees that were then about 18 m tall. Tree ring counts made at the same time showed that trees dated from about 1908.

The site of the Geescroft Wilderness was cropped for about 30 years with beans, followed by 4 years bare fallow and 3 years clover immediately prior to abandonment in 1885 (Jenkinson, 1970). The wilderness, which was fenced off from stock and surrounded on three sides by high hedges with a good many trees, occupies an area of about 0.8 ha. After 10 years little recolonisation had occurred, the vegetation consisted of a few tall, tufted grasses and a few small *Quercus* seedlings near one of the large oaks (Lawes, 1895). By 1913-14 Geescroft remained a desolate area densely covered with tufted hair-grass (*Deschampsia cespitosa*) dotted with a few small trees and shrubs; these were gorse (*Ulex europaeus*), *Corylus avellana*, *Q. robur*, *A. campestre*, *S. nigra*, elm (*Ulmus minor*), *F. excelsior*, *Crataegus monogyna*, *Prunus* sp., *Rubus* sp. and *Rosa* sp. (Brenchley and Adam, 1915). During the next 40+ years the area recolonised to a woodland which, by 1957, consisted largely of *U. minor*, *F. excelsior* and *Q. robur*.

Although there has been no systematic surveying at either site, the information available suggests that the pattern of recolonisation differed between the two sites. At Broadbalk the vegetation present during the first 20 years after abandonment was dominated by herbaceous species but after 30 years the wilderness had developed into dense thicket with some plants characteristic of a woodland ground flora. In contrast, recolonisation at Geescroft was slower, with woodland taking 30–70 years to develop. Reasons for this are unclear. The sites are 1.3 km apart with both having a leached brown soil and a loamy surface layer overlaying clay-with-flints. The sites had different crop histories and apparently Geescroft had more rabbits, but Brenchley and Adam (1915) suggest that a difference in drainage was most important. It is probable that the abundance of *D. cespitosa*, a grass characteristic of wet sites, would severely inhibit recolonisation by providing strong competition to tree seedlings.

Natural colonisation is also seen as a method by which woodland can be established on derelict urban and industrial land, and in a recent survey it was found to have created 35% of the woodland present in the Borough of Sandwell, an urban area in the West Midlands (Anon, 1991b). Information available suggests that the between-site variation in time scales for establishment is similar to that for agricultural land. The spread of woody plants onto coal-pit heaps is generally related to their age with few species at low abundance on heaps 1–20 years old and with more species at greater abundance on heaps over 50 years old, although there is considerable variation between heaps of similar age (Hall, 1957). These differences were related not only to environmental factors such as soil pH but also to the proximity of seed trees. Other types of spoil colonise equally slowly (Bradshaw and Chadwick, 1983). Recent evidence suggests that if colonisation does not occur quickly it will take many years to develop. A survey of 47 derelict sites more than 1 ha in area, located in central and west England, showed that only 30% of sample quadrats (4 × 4 m in size, 13 per hectare) were colonised by one or more woody plants within 10 years and that this figure remained more or less constant until sites had been abandoned for more than 35 years (S. J. Hodge, personal communication).

During the early stages of colonisation, which may last several decades, the species invading will depend on the surrounding sources of seed. Very often these will be species with lightweight windblown seed such as willow (*Salix* spp.) and *Betula* spp., or bird-dispersed species such as *C. monogyna*, which were the three most frequently found species in Hodge's survey. Heavier keys of *F. excelsior* and *A. pseudoplatanus* may spread 1–200 m from the parent, whereas seeds of *Quercus* spp. and *Fagus sylvatica* may be carried a few kilometres by birds (Johnson and Adkisson, 1985). The communities which develop in this way can be species poor, for example there were only four species of tree recorded on 32-year-old iron-ore spoil banks (Leisman, 1957), and it is often found that urban colonisation is dominated by *Salix* spp. and *Betula* spp.

Any difficulties experienced with the use of natural colonisation will be similar to those for natural regeneration. Proximity to seed sources is likely to be especially important, whereas browsing damage may be less of a problem in urban areas. Success will depend on many factors and may vary widely with location of the site: in agricultural areas where soils are good, rapid growth of weeds is likely to inhibit colonisation, but on derelict urban sites the soil may be poor enough to prevent growth of a vigorous ground cover and colonisation may occur if seeds are available. If colonisation fails and a closed, permanent ground cover develops then it may be necessary to carry out some kind of vegetation management (such as herbicide treatment or scarification) and encourage seed dispersal by provision of bird perches (McClanahan and Wolfe, 1993; McDonnell and Stiles, 1983; Robinson and Handel, 1993).

The process of natural colonisation is very unpredictable and it may take many years to achieve a woodland environment which is dominated by trees. If there is a need to establish woodland of a known type or within a defined time period then colonisation is probably unsatisfactory and the conclusion of Ellis (1745), that colonisation should be used to augment woodlands, remains true and is a basic tenet of planting designs recommended by Rodwell and Patterson (1994).

Direct seeding

Direct seeding is the process by which woodlands are established on previously unwooded sites by the artificial sowing of tree seed directly where trees are to be grown. One of the earliest references to the use of direct seeding as a method of establishing a crop of trees was when the German, Henry VII, in 1309 ordered the reforestation of an area by the sowing of tree seed (Fernow, 1913). According to James (1981) the practice of direct sowing began in England in the latter part of the 15th century, but the earliest reference that he could find was by Roger Traverner, in a survey dated 1565, referring to the sowing of acorns and beech mast. Many writers in the 17th and 18th century describe a myriad of different methods of direct sowing, including Evelyn (1670), Ellis (1745), Cook (1679) and Langley (1728), although the original source of these descriptions and the experiences on which the recommendations are based are sometimes obscure.

There were two basic methods of sowing seed of broadleaved species for this purpose. The first which was applicable to unplanted land, consisted of sowing them with, or shortly after, a cereal crop such as wheat. The second was by planting a small number of acorns in a hole, a process known as dibbling, which was normally used when replanting old woodland. The earliest record of the first method is given by Tusser (quoted in James, 1981) who advised:

> Sow acornes, ye owners that timber do love
> Sow hay and rie with them, the better to prove
> If cattle or coney may enter the crop
> Young oak is in danger of losing his top.

The technique was not confined to oak and Mortimer (1708) recommended its use with ash keys:

> But if you would make a wood of them (ash keys) at once, dig or plough up a parcel of land and prepare it as for corn; only if you plough it give it a summers fallowing to kill and rot the turf, ploughing of it as deep as you can, and with your corn, especially oats, sow your ash keys, and at harvest taking off your crop of corn the next spring you will find it covered with young ashes, which by reason of their small growth the first year should be kept well weeded and well secured from cattle (domestic stock), who are very desirous of cropping them, the second year they will strike root, and quickly surmount what impediments they may meet.

In addition, an example in Evelyn (1670) recommends the sowing of acorns with gorse (*Ulex* spp.) which supposedly reduced the requirement to fence and protect the crop from mammals.

The practice of dibbling was common and an example from the Forest of Dean is cited by James (1981):

> The method of planting is, first, to mark out the ground; then taking off about a foot square of turf, to set two or three acorns with a setting-pin; afterwards to invert the turf upon them and, by way of raising a fence against hares and rabbits, to plant two or three strong whitethorn sets around.

The problems of direct seeding (predation by mice and birds, variable germination and prolific weed growth) were slowly recognised by practitioners. This led to the practice, at the beginning of the 17th century, of sowing seed into small pieces of ground where predation and weed growth could be controlled. In turn this led to the development of forest nurseries and the now common practice of raising seedlings in a nursery before planting out in the field.

A more modern description of direct sowing can be found in Schlich (1904) from which we derive the following guidance:

- Hardy species with large seeds are better suited to direct sowing compared with delicate or expensive seeds.
- Use of seeds which have shown good germination in tests is essential. The criteria of good seed are defined in terms of a lowest acceptable percentage germination (Table 8.2).
- Sow sufficient seed; the amount varies with species. The quantities of seed recommended have been converted to metric units in Table 8.2. In all cases the number of seeds recommended by Schlich are well above the minimum of 100 000 seeds per hectare recommended for broadleaves by Evans (1984): those for *Betula* spp. and alder (*Alnus glutinosa*) are extremely high and could possibly be reduced tenfold. The present cost of the seed to cover 1 hectare has also been calculated (Table 8.2) and varies from £255 for *Acer pseudoplatanus* to £4080 for *Betula* spp. In all cases, except *Fraxinus excelsior* and *A. pseudoplatanus*, the cost of the seed element of Schlich's recommendations at present prices would be in excess of the cost of buying plants for an afforestation scheme using 2000 trees per hectare. One method Schlich recommends for reducing the quantity of seed used is by sowing seed in strips or furrows; the use of trenches could reduce the seed requirement by 30% and the use of pits or holes (dibbling) by 75%.

- Schlich recommends that the best time to sow seed is in April or May but this may not be true if the seeds being sown require a period of cold to break their dormancy, for example *Fraxinus excelsior, Fagus sylvatica*, hornbeam (*Carpinus betulus*) or *A. pseudoplatanus*.
- Irrespective of the method of sowing, whether by broadcast, trench or dibbling, the seed must be covered by soil to protect against predation, desiccation and sudden temperature change. The optimum depth of soil coverage varies with species (Table 8.3). The use of ploughing is recommended to cover oak but a number of other methods such as harrowing, raking, hoeing or covering with fine soil can be successful for a broad range of species.

Table 8.2. Schlich's (1904) recommendations and present costs of direct sowing

Species	% Germinable[a]	Sowing density		Cost of seed 1993[b]	
		kg ha⁻¹	No. seeds ha⁻¹ (000's)	£ kg⁻¹	£ ha⁻¹
Pedunculate oak	65	616	177	3.25	2000
Sessile oak	65	616	217	3.25	2000
Beech	50	168	741	12.50	2100
Sycamore	55	34	370	7.50	255
Ash	65	39	561	7.50	293
Hornbeam	65	39	1296	20.00	780
Common alder	30	17	11 111	17.00	1190
Birch	20	34	59 260	120.00	4080
Norway spruce	75	11	1605	150.00	1650
Larch	35	16	2420	250.00	4000
Scots pine	70	7	1111	140.00	980

[a] Schlich definition of germinable is 'good and clean'.
[b] From Forest Enterprise seed catalogue 1993–94.

Table 8.3. Recommended covering depths for seed based on Schlich (1904)

Species	Depth (cm)
Pedunculate oak	4
Beech	2
Sycamore	1.5
Scots pine and spruce	1.5
Common alder	1
Larch	1

The continued interest in direct sowing led researchers in the Forestry Commission to establish a number of field experiments; however, results have been very variable (Stevens *et al.*, 1990) and the probability of failure is high. There are three main reasons for this high risk of failure.

- The emergence of tree seeds is slow and this gives animal, bird and insect predators time to inflict heavy losses on the seeds sown (Evans, 1984).

● The emergence of tree seeds is unpredictable, and even laboratory tested seeds which are known to be alive and capable of germinating under optimum conditions often fail to germinate in the field (Gosling, 1987).

● The slow early growth of the trees relative to other annual and perennial species means that woody plants cannot reliably compete against other vegetation (Davies, 1987).

These difficulties have led to the conclusion that traditional establishment techniques, of raising seed in nurseries and then transplanting, are superior to direct sowing (Stevens *et al.*, 1990). Despite the perceived problems of direct sowing some contemporaries believe that the technique has advantages over traditional methods and even claimed direct sowing costs can be 50% less than planting (Anon, 1988). In addition Luke (1984) has suggested that direct seeding has advantages over traditional methods for establishing trees on difficult sites, e.g. reclamation sites, steep slopes and areas with awkward access. He has also claimed that the appearance of vegetation arising from direct seeding is more natural than that created by planting nursery stock.

The use of direct seeding of tree seed in combination with agricultural crops, similar in many ways to the methods described above by Tusser (in James, 1981) and Mortimer (1708), is currently being practised commercially on a number of sites in south-west England (Plates 8.1, 8.2 and 8.3). Early results from sites being monitored by Forestry Authority Research Division are given in Table 8.4 and promising seedling densities of 6710 ha^{-1} and 16 670 ha^{-1} have been attained on two sites. However, in an effort to minimise costs the seedlings have little effective protection from mammals and high mortalities are expected after removal of the agricultural cover crop. In addition, practicable methods of controlling the weeds in close proximity to seedlings have not been developed. It is too early to say if the ingenuity of modern foresters can overcome these constraints which were the main difficulties encountered in the use of direct sowing in the 18th and 19th centuries.

Plate 8.1. An agricultural site sown with a mixture of wheat and tree seed of oak (*Quercus* spp.), sycamore (*Acer pseudoplatanus*), ash (*Fraxinus excelsior*), beech (*Fagus sylvatica*), Norway maple (*Acer platanoides*), and hazel (*Corylus avellana*). For results see site 1, Table 8.4. (D. Rogers, The Forestry Authority)

Plate 8.2. Precision sowing of oak (*Quercus* spp.). See site 1, Table 8.4. (D. Rogers, The Forestry Authority)

Plate 8.3. Seedling oak (*Quercus* spp.) among a wheat crop. See site 1, Table 8.4. (D. Rogers, The Forestry Authority)

ESTABLISHMENT OF WOODLANDS BY PLANTING

Since the mid-17th century the main method of establishing woodlands in Britain has been by planting and there are reliable methods for establishing the woody species used in forestry (Davies, 1987; Hart, 1991; Hibberd, 1991; James, 1989; Kerr and Evans, 1993; Potter, 1991). In the past, foresters have tried to match the site with the species that is likely to be the most productive, often with little thought for the appearance of the woodland and its role in the environment on either a site or landscape scale. However, increased awareness of the importance of woodlands in the landscape has stimulated improvements to woodland design (Bell, Chapter 3) and ideas for the choice and spatial arrangement of species (Rodwell and Patterson, 1994; Rodwell and Patterson, Chapter 5). While the size and shape of woodlands are relatively easy to plan and manage, complex mixtures and planting patterns will be more difficult to design, establish and manage than simple, single species stands.

Table 8.4. Forestry Authority monitoring of direct seeding: sowing densities and number of seedlings present in autumn one or two years after sowing

Species	Site 1[a]		Site 2[b]	
	Seed sowing density (No. ha^{-1})	Seedlings present after 1 year: as % of sowing density (%)	Seed sowing density (No. ha^{-1})	Seedlings present after 2 years: as % of sowing density (%)
Sycamore	25 000	7.6[c]	30 000	14.6
Norway maple	15 600	—	20 000	—
Field maple	5500	0.2	—	22.5
Ash	35 100	27.0	40 000	—
Oak	10 000	4.1	—	—
Wild cherry	9800	1.5	—	—
Hazel	2300			3.7
Beech	—	—	10 000	—

[a] Precision sown oak, other species broadcast sown and harrowed.
[b] All species broadcast sown and harrowed.
[c] Aggregated results as plants assessed after leaf fall. Results were combined as plants were difficult to identify in leafless state.

Canopy establishment

The establishment of a woodland canopy is one of the most noticeable features of woodland creation, providing, for example, visual amenity, habitats for wildlife and shady conditions suitable for the woodland ground flora. The speed with which a canopy is created will vary, depending on site, species and management but it will generally be achieved most rapidly with high planting densities. At 2 × 2 m spacing, well-maintained wild cherry (*Prunus avium*) and oak grown in treeshelters (Potter, 1991) can close canopy after 4 and 6 years, respectively. Observations of tree growth at 3 m spacing on fertile agricultural soils suggest that small-leaved lime (*Tilia cordata*), *P. avium*, *A. pseudoplatanus* and *A. campestre* may take 8 years to close canopy (I. Willoughby, personal communication). Factors which reduce growth rate, such as low site fertility, poor silvicultural practice and browsing damage, will extend the time to canopy closure. For slow growing species such as *Quercus* spp. and *F. sylvatica* at 3 × 3 m it may take 15 years even on good fertile soils. The most appropriate initial spacing has recently been the subject of much debate and should be guided by the objectives of the planting scheme and the nature of the site to be planted (Kerr, 1993). However, many planting schemes are dominated by financial considerations and the constraints of any grants available. The Woodland Grant Scheme which aids planting and management of woodlands in Britain (Anon, 1991a), allows a minimum planting density of 1100 stems per hectare (3 × 3 m spacing). This is neither good for creation of a closed canopy nor the production of high quality timber trees; 2500 stems per hectare (2 × 2 m spacing) is a better option for both purposes. Rodwell and Patterson (1994; and also Chapter 5) give a number of innovative ideas for planting patterns within woodlands which include variable spacing between trees, varying the sizes of clumps and also the gaps between them. While these will probably be more expensive to establish and manage than uniform planting, and difficult to include in small planting schemes, they are likely to have a pronounced effect on the structural diversity of the woodland. However, use of these ideas may be constrained by other needs, for example, the minimum clump size required to produce shaded conditions suitable for woodland herbs is about 0.15 ha (Buckley and Knight, 1989). In addition widely spaced trees or clumps will create sheltered woodland glades which provide a favourable habitat for deer which may thwart the introduction of a varied shrub and ground flora into the surrounding woodland (Gill and Gurnell, Chapter 13; Packham *et al.*, Chapter 9).

Planting patterns

Planting gives exciting opportunities for choice of species, and guidance for this is given in Rodwell and Patterson (1994), but the temptation to use complicated mixtures should generally be resisted as they may fail or be difficult to manage. Intimate mixtures are the most difficult to manage and should only consist of compatible species. Although Rodwell and Patterson (1994) have suggested suitable mixtures, their success is likely to vary with site, proportion of each species in the mixture, initial densities and growth rate. Mixtures are more likely to succeed with individual species planted in groups, the size of which may vary with species and depend on the crown projection of the mature tree. The recommended stocking and estimated crown projection of six common woodland species at the end of their rotation are shown in Table 8.5. The basic planting unit would be the group rather than the tree with variety created by planting at uniform spacing within each group but at different spacing between groups and planting the groups in planned or random positions as suggested by Helliwell (1993).

Table 8.5. Size of planted group for different broadleaved trees estimated from crown projection of final crop trees at the end of the rotation

Species	Normal rotation age (yr)	Final crop stocking (stems ha^{-1})	Crown projection (m^2)	Group size (m × m)
Ash	65–75	120–150	67–83	8 × 8
Beech	95–140	100–120	83–100	10 × 10
Cherry	50–70	140–160	63–71	8 × 8
Oak	120–160	60–90	111–167	12 × 12
Sycamore	60–70	140–170	59–71	8 × 8
Sweet chestnut	60–70	100–190	53–100	10 × 10

Establishing woodlands by planting is a well-practised and reliable technique which has advantages over natural colonisation and direct seeding as it allows good control over the mixture of species present, their density and position on the site and, with good management, establishment of trees is rapid. However, all methods of establishment provide no direct control over the development of a varied woodland ground flora. There may be differences in the suitability of sites for the establishment of trees and ground flora. For example, sites with good agricultural soils will encourage the rapid growth of properly maintained trees but their fertility may present problems for the establishment of small, slow growing herbs. At present there is great interest in the development of methods to improve the diversity of the ground flora in recently planted secondary woodlands (Packham *et al.*, Chapter 9).

MANAGEMENT

The management required to create a new woodland will be site specific and depends not only on the physical, environmental and biological characteristics of the site, but also on the objectives for the woodland and the method chosen for its creation. In most cases neglect will not be a good option as positive management will encourage the development of desired woodland characteristics. For example, very dense areas of colonisation will rapidly produce a canopy providing shady conditions but unless these are thinned the trees which survive may be very poor and cast too dense a shade for development of a good ground flora. The silvicultural procedures which may be necessary to establish a woodland within 15 years are shown in Table 8.6. With careful planning and use of good silvicultural practice most of the input for planted woodlands occurs in the first 5 years, with maintenance only in years 5–15. In contrast, natural colonisation and direct seeding are likely to have a continuing need for management beyond year 5. If direct seeding has been successful, with good germination and early survival, the most important operation will be respacing over some or all of the site. Although dense areas of natural colonisation may need thinning, the most probable continuing operations for this method of establishment are ground preparation and vegetation management, which will maintain conditions suitable for invasion by seed.

CONCLUSIONS

The relative costs of each method of establishment for any site are very difficult to determine as they depend on the degree of success with early establishment. Planting is the most predictable with most costs occurring early. It may appear that these will be much higher than those for natural colonisation but this is not necessarily true if colonisation is either slow and patchy with some planting to fill in gaps, or very dense with the need for thinning. Conversely the initial high cost of planting may be set against the earlier visual impact it can achieve, particularly in woodlands planned for their amenity value. Whichever method of establishment is chosen it is important that it can fulfil the objectives for woodland creation within the desired timescale, that the potential costs of establishment are thoroughly assessed, and that proper provision is made for early management.

Table 8.6. Operations which may be needed to ensure rapid woodland establishment within 15 years

		Natural colonisation	Direct seeding	Planting
Years 1–5	Deer/rabbit fencing[a]	**	**	**
	Individual tree protection[b]	*	–	**
	Plants	–	–	**
	Ground preparation	**	**	*
	Vegetation management	**	**	**
Years 5–10	Individual protection	*	*	–
	Plants	*	*	–
	Ground preparation	*	–	–
	Vegetation management	*	*	–
	Thinning	*	*	–
	Maintenance	*	*	*
Years 10–15	Individual protection	*	*	–
	Plants	*	*	–
	Ground preparation	*	–	–
	Vegetation management	*	*	–
	Thinning	*	*	–
	Maintenance	*	*	*

** Probable requirement.
* Possible requirement.
– No requirement.
[a] For planting, will not fence if using tree shelters.
[b] Tree shelters, vole and rabbit guards, etc.

REFERENCES

Anon (1988). Saving your money with direct seeding. *Horticulture Week*, **203**(6), 32.

Anon (1990). *The Management of Broadleaf Woodland in the Forest of Dean*. Forestry Commission, West England Conservancy, Coleford, 15pp.

Anon (1991a). *Forestry Policy for Great Britain*. Forestry Commission, Edinburgh, 13pp.

Anon (1991b). Sandwell's Urban Forest. Black Country Urban Forestry Unit, Sandwell, 14 pp.

Benezit, J.J.(1991). Soils and natural regeneration in Normandy. *Quarterly Journal of Forestry*, **85**, 30–36.

Bradshaw, A.D. and Chadwick, M.J. (1983). *The Restoration of Land*. Blackwell Scientific Publications, Oxford, 317 pp.

Brenchley, W.E. and Adam, H. (1915). Recolonization of cultivated land allowed to revert to natural conditions. *Journal of Ecology*, **3**, 193–210.

Buckley, G.P. and Knight, D.G. (1989). The feasibility of woodland reconstruction. In: Buckley, G.P. (ed.). *Biological Habitat Reconstruction*. Belhaven Press, London, 171–188.

Collin, P.H. (1992). *Dictionary of Ecology and the Environment*. Collins, London, 236pp.

Cook, M. (1679). *The Manner of Raising, Ordering and Improving Forest and Fruit Trees*. Peter Parker, London, 204pp.

Davies, R.J. (1987). *Trees and Weeds*. Forestry Commission Handbook 2. HMSO, London, 36pp.

Ellis, W. (1745). *The Timber-Tree Improved*, 4th edn. T. Osborne and M. Cooper, London, 207pp.

Evans, J. (1984). *Silviculture of Broadleaved Woodland*. Forestry Commission Bulletin 62. HMSO, London, 232pp.

Evans, J. (1988). *Natural Regeneration of Broadleaves*. Forestry Commission Bulletin 78. HMSO, London, 46pp.

Evelyn, J. (1670). *Sylva or a Discourse of Forest Trees*, vol. 1, 4th edn. Doubleday & Co., York, 335pp.

Fernow, B.E. (1913). *A Brief History of Forestry in Europe, the United States and Other Countries.* University Press, Toronto, 506pp.

Ford-Robertson, F.C. (ed.) (1971). *Terminology of Forest Science, Technology, Practice and Products.* IUFRO/Society of American Foresters, Washington DC, 97pp.

Gill, D.S. and Marks, P.L. (1991). Tree and shrub seedling colonization of old fields in central New York. *Ecological Monographs*, **61**, 183–205.

Gosling, P.G. (1987). Dormant tree seeds can exhibit similar properties to seed of low vigour. In: Patch, D. (ed.). *Advances in Practical Arboriculture.* Forestry Commission Bulletin 65. HMSO, London, 28–31.

Hall, I.G. (1957). The ecology of disused pit heaps. *Journal of Ecology*, **45**, 689–720.

Hanks, J.P. (1971). Secondary succession and soils on the inner coastal plain of New Jersey. *Bulletin of the Torrey Botanical Club*, **98**, 315–321.

Harrison, J.S. and Werner, P.A. (1984). Colonization by oak seedlings into a heterogeneous successional habitat. *Canadian Journal of Botany*, **62**, 559–563.

Hart, C. (1991). *Practical Forestry for the Agent and Surveyor*, 3rd edn. Alan Sutton, Stroud, 658pp.

Helliwell, R. (1993). The patterns of nature. *Landscape Design*, **220**, 18–20.

Hibberd, B.G. (1991). *Forestry Practice.* Forestry Commission Handbook 6. HMSO, London, 239pp.

James, N.D.G. (1981). *A History of English Forestry.* Blackwell, Oxford, 339pp.

James, N.D.G. (1989). *The Forester's Companion*, 4th edn. Basil Blackwell, Oxford, 310pp.

Jenkinson, D.S. (1970). The Accumulation of Organic Matter in Soil Left Uncultivated. Report for the Rothamsted Research Station for 1970, Part 2, 113–137.

Johnson, W.C. and Adkisson, C.S. (1985). Dispersal of beech nuts by blue jays in fragmented landscapes. *American Midland Naturalist*, **113**, 319–324.

Kerr, G. (1993). Robust woodland establishment. *Quarterly Journal of Forestry*, **87**, 302–304.

Kerr, G. and Evans, J. (1993). *Growing Broadleaves for Timber.* Forestry Commission Handbook 9. HMSO, London, 104pp.

Langley, B. (1728). *A Sure Method of Improving Estates.* Francis Clay, London, 274 pp.

Lawes, J.B. (1895). Upon some properties of soils. *Agricultural Students Gazette*, 7, 64–72.

Leisman, G.A. (1957). A vegetation and soil chronosequence on the Mesabi iron range spoil banks, Minnesota. *Ecological Monographs*, **27**, 221–245.

Luke, A. (1984). Trees: naturally. *Mineral Planning*, **20**, 40–43.

McClanahan, T.R. and Wolfe, R.W. (1993). Accelerating forest succession in a fragmented landscape: the role of birds and perches. *Conservation Biology*, **7**, 279–288.

McDonnell, M.J. and Stiles, E.W. (1983). The structural complexity of old field vegetation and the recruitment of bird dispersed plant species. *Oecologia*, **56**, 109–116.

Matthews, J.D. (1989). *Silvicultural Systems.* Oxford University Press, Oxford, 284pp.

Maycock, P.F. and Guzikowa, M. (1984). Flora and vegetation of an old field community at Erindale, southern Ontario. *Canadian Journal of Botany*, **62**, 2193–2207.

Mellanby, K. (1968). The effects of some mammals and birds on regeneration of oak. *Journal of Applied Ecology*, **5**, 359–366.

Mortimer, J. (1708). *The Whole Art of Husbandry*, 2nd edn. Mortlock, London, 632pp.

Pickett, S.T.A. (1982). Population patterns through twenty years of oldfield succession. *Vegetatio*, **49**, 45–59.

Potter, M.J. (1991). *Treeshelters.* Forestry Commission Handbook 7. HMSO, London, 48pp.

Raup, H.M. (1940). Old field forests of south eastern New England. *Journal of the Arnold Arboretum*, **21**, 266–273.

Robinson, G.R. and Handel, S.N. (1993). Forest restoration on a closed landfill: rapid addition of new species by bird dispersal. *Conservation Biology*, **7**, 271–278.

Rodwell, J. and Patterson, G. (1994). *Creating New Native Woodlands.* Forestry Commission Bulletin 112, HMSO, London, 88pp.

Schlich, W. (1904). *Schlich's Manual of Forestry*, vol. 2, *Silviculture*, 3rd edn. Bradbury, Agnew and Co., London, 393pp.

Smith, D.M. (1986). *The Practice of Silviculture*, 8th edn. Wiley, New York, 527pp.

Stace, C.A. (1991). *New Flora of the British Isles.* Cambridge University Press, Cambridge, 1226pp.

Stevens, F.R.W., Thompson, D.A. and Gosling, P.G. (1990). *Research Experience in Direct Sowing for Lowland Plantation Establishment*. Forestry Commission Research Information Note 184. Forestry Commission, Edinburgh.

Teissier du Cros, E. (1987). The French approach to broadleaved silviculture. *Irish Forestry*, **44**, 116–126.

Watt, A.S. (1924). On the ecology of British beech woods with special reference to their regeneration. Part II. The development and structure of beech communities on the Sussex Downs. *Journal of Ecology*, **12**, 145–204.

Witts, K.J. (1964). Broadbalk Wilderness Flora. Report for the Rothamsted Research Station for 1964, 219–222.

9   Introduction of Plants and Manipulation of Field Layer Vegetation

J. R. PACKHAM, E. V. J. COHN, P. MILLETT and I. C. TRUEMAN

INTRODUCTION

These are exciting times for woodland ecologists. Knowledge of semi-natural British woodlands has never expanded so rapidly as it has in the last two decades, nor has it previously been employed to consider such matters as 'Creating New Native Woodlands' (Rodwell and Patterson, 1994). We are, moreover, gradually acquiring more and more experience of woodland reconstruction and habitat creation.

Buckley and Knight (1989) complete their survey of the feasibility of woodland reconstruction by considering experience gained in moving Biggins Wood, Kent as a consequence of construction of the Channel Tunnel Terminal at Folkestone. Trials in anticipation of a similar transplantation of the whole of Darenth Wood, Kent (Down and Morton, 1989) have also yielded valuable experience. In North America, despite the present wide knowledge of field layer plants, the main thrust has been directed towards establishing tree and shrub layers (Ashby, 1987).

The problems encountered in creating diverse woodland communities, from bare fields or by the transformation of low quality woodlands, are very similar but involve the additional difficulty of obtaining suitable plant material of proven quality on a commercial scale (Brown, 1989). There are at least three major reasons why it is worth persisting with such attempts:

- The creation of such woodland can increase the quality of life of city dwellers living adjacent to suitable sites.
- Attractive field layers in such woodlands tend to attract visitors away from relatively rare ancient semi-natural woodlands of high scientific and conservation value.
- Experience gained during the manipulation or creation of diverse field layers is likely to increase our understanding of how woodlands function normally; for example Ashby (1987) has suggested that forest restoration is an opportunity to test ideas about how woodlands develop, and to explore our understanding of natural succession.

Our approach (Packham and Cohn, 1990, see Table 9.1; Cohn and Packham, 1993) has been based on a knowledge of the individual species, e.g. Ellenberg (1991), their

The Ecology of Woodland Creation. Edited by Richard Ferris-Kaan.

occurrences and interactions in nature, e.g. Grubb (1977), and a practical application of C-S-R theory (Grime, 1979; Grime et al., 1988; Hodgson, 1989). Knowledge of British woodland vegetation, both on a national (Peterken, 1981; Rodwell, 1991) and a more local scale (e.g. Trueman et al., 1985) is now well established. Information regarding the distribution, physiology and behaviour of particular species is also increasing, notably through biological flora accounts, but also through other studies some of which are directly concerned with habitat creation (e.g. Francis et al., 1992a; Fu, 1993). Studies of the effects of tree leaf litter upon the emergence of the aerial organs of woodland plants are particularly relevant (Salisbury, 1916; Sydes and Grime, 1981).

Table 9.1. Important characteristics of woodland field layer plants. Source: Packham & Cohn (1990), Arboricultural Journal, reproduced by permission of A.B. Academic Publishers

a.	*Life forms*	dwarf shrubs (including nanophanerophytes with perennating buds 25 cm to 2 m above ground) chamaephytes (perennating buds up to 25 cm above soil level) hemicryptophytes (perennating buds at soil level) geophytes (root stocks, rhizomes, bulbs, corms) helophytes (marsh plants) hydrophytes (water plants: occur in woodland ponds) therophytes (endure unfavourable season as seeds)
b.	*Duration of life*	annual, biennial, perennial
c.	*Means of reproduction*	seeds (? ability to form seed banks), stolons, runners, rhizomes, bulbils. Seed and fruit dispersal: animal, wind, explosive mechanisms
d.	*Seed dormancy*	period that can remain viable, germination requirements
e.	*Importance of any mycorrhizal associations*	
f.	*Influence of soil factors*	soil pH, requirements for bases e.g. Ca^{2+}, Mg^{2+}, K^+, N, P requirements. Vulnerability to Al^{3+}, Fe^{2+}. Drainage, soil texture and aeration
g.	*Reaction to shade*	sun/shade plant? Shade evader or shade tolerator?
h.	*Phenology*	(seasonal pattern of development) vegetative growth, flowering, fruiting
i.	*Relative growth rate (RGR) and competitive ability*	
j.	*Characteristic habitat*	heavy/light shade; frequently/seldom disturbed; under main canopy/at glade margin/along paths; by streams/in ponds
k.	*Vulnerability to biotic factors*	grazing, trampling, insect/slug damage; fungal, viral and bacterial disease
l.	*Any special features*	e.g. poisonous to stock; unusual vulnerability to late frosts, drought, or flower picking

The soil is one of the most fundamental factors involved in the creation of new broadleaved woodland (Moffat and Buckley, Chapter 6), and where highly fertile soil is combined with relatively light shading, as in many plantations at Milton Keynes, Buckinghamshire, non-introduced weed species may reach a height of up to 2 m (Seed, 1993). Soil quality, the nature of the plant litter and of mulch residues, will influence both the invertebrate populations and the microbial ecology of the soil itself (Harris and Hill, Chapter 7). The choice of tree species, the density at which young trees are planted, the nature of their rooting systems, phenology and leaf litter, together with the method of establishment and the shade they cast, are most important (Harmer and Kerr, Chapter 8). Cherry (*Prunus avium*)†, for example, is a major element in the woodland at Nedge Hill, Telford New Town, Shropshire, which has been used in some of our experiments. Although planted only 15–20 years ago, these trees are now seeding and suckering prolifically and the woodland floor has so many of their saplings that establishment of a satisfactory herb layer is rendered difficult.

This chapter is mainly concerned with the results of experimental work at the University of Wolverhampton Field Station, and with larger scale establishment trials in the urban woodlands of Wolverhampton, West Midlands. Decisions regarding choice of trees and shrubs came before field layer experiments were undertaken. Should such choices in future be bound entirely by considerations of visual amenity or, at the other extreme, should an attempt be made to create simulacra of natural communities? In the latter case, choice of species is likely to be informed by reference to local conditions and to communities described in the National Vegetation Classification (Rodwell and Patterson, Chapter 5).

Habitat creation and conservation

In our view, the objectives of habitat creation schemes should be clearly defined at the outset and adhere to a sensible ethical code of practice (Packham and Cohn, 1990). Habitat creation frequently involves the use of horticultural techniques, but its essential aim should be to introduce suitable species in such a way that they will hold their own in a developing community rather than being protected by intensive later operations such as hand weeding. Wherever possible planting stocks should be created from local races and care should be taken to avoid future confusion by planting rarities (see Spencer, Chapter 1), a situation which would be further compounded by a failure to maintain adequate records. By and large, successful habitat creation schemes are unlikely to achieve stable or easily maintained simulacra of complex semi-natural communities, and should avoid using taxa which will cause unnecessary complications for botanical recorders. Habitat creation should not be used in sweeping 'tidying up' operations of the type that may destroy unusual communities of great intrinsic interest which not infrequently develop in towns (Gilbert, 1989) and on old mineral extraction sites.

The word 'gardening' is frequently mentioned by critics of habitat creation. The essential difference between what we are trying to do in habitat creation and what the Dutch are doing in creating and maintaining imitations of various habitat types (Ruff, 1979) is that, by and large, our intention is to introduce species in such a way that they will hold their own in a developing community rather than being protected by intensive later operations such as weeding, which is an essential feature of most gardens.

---

† Nomenclature follows Clapham, A.R., Tutin, T.G. and Moore, D.M. (1987) *Flora of the British Isles.* Third Edition, Cambridge University Press, Cambridge.

Species strategies and habitat creation

Figure 9.1 is a C-S-R ordination diagram (*sensu* Grime, 1979) for plants of the woodland field layer in which the positions occupied by particular species are related to the strategies they exhibit. Rosebay willowherb (*Chamaenerion angustifolium*), creeping soft-grass (*Holcus mollis*), bracken (*Pteridium aquilinum*) and bramble (*Rubus fruticosus* agg.) are all familiar as very strong competitors, while nettle (*Urtica dioica*) is favoured by the high phosphorus levels commonly found in urban woodlands. Chickweed (*Stellaria media*) is a true ruderal commonly associated with disturbance, and nipplewort (*Lapsana communis*) has similar tendencies. Wood sorrel (*Oxalis acetosella*), on the other hand, is a species particularly well suited to the stresses imposed by heavy shading (Packham, 1978; Packham and Willis, 1976). In this it resembles another true *shade-tolerator,* yellow archangel (*Lamiastrum galeobdolon*) (Packham, 1983; Packham and Willis, 1982).

Bluebell (*Hyacinthoides non-scripta*), found close to *O. acetosella* on the ordination diagram, survives the dense shade of deciduous woodlands as a *shade evader* whose active vegetative period precedes the flushing of the tree canopy. The responses of this species to shade, edaphic conditions and other factors are also well known (Blackman and Rutter, 1954; Knight, 1964; Grabham and Packham, 1983).

Habitat creation in urban woodlands and on set-aside land involves a wide variety of substrates and the creation of vegetation whose planned duration of life may vary from a few years, in the case of a city centre site covered with brick rubble, to long term in the case of permanent recreational land. The objective is to create or manipulate the woodland field layer so that attractive field layer species are established or encouraged, and species perceived to be unattractive weeds are eliminated or controlled.

In British woodlands shade is the major cause of stress in field layer plants and this factor can be used to manipulate the vegetation on a long-term basis. Soil nutrient levels must also be taken into account: high levels of nitrogen and phosphorus will encourage *U. dioica* and other tall ruderals, but use may be made of the fact that several visually attractive species such as leopard's bane (*Doronicum pardalianches*), *L. galeobdolon* and red campion (*Silene dioica*) will grow under similar conditions. As little can be done to reduce soil fertility in the short term, the desired community type and the growth regime to be employed must be selected with this parameter in mind.

Our experience has led to the conclusion that in many urban situations persistent weeds are such a major problem that firm steps must be taken to control them from the beginning. Use of herbicides and rotavation provides only an initial check on the growth of many ruderals, and for this reason much of our work has been concerned with detailed assessments of the value of mulches in woodland weed control.

OUTLINE OF EXPERIMENTAL PROGRAMME

The first experiment in the current programme was begun in October 1989 at the Compton Field Station, University of Wolverhampton. It was designed to investigate the influence of four surface litter treatments on the establishment and reproduction of five woodland herbs, three species being established as transplants and two from seed. Subsequent work has been concerned with establishing field layer communities from

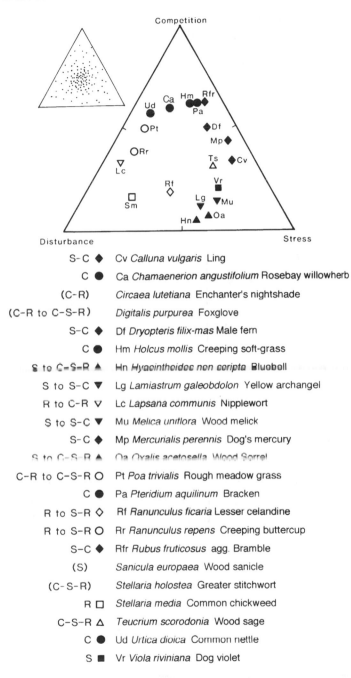

Figure 9.1. C-S-R ordination diagram of woodland field layer species. This describes the various equilibria between competition, stress and disturbance in vegetation. The points shown for each species indicate the position of its maximum percentage occurrence in a matrix of vegetation types classified according to the strategies of the component species in Grime *et al.* (1988). Where points are not provided, the data for the species concerned are insufficiently clear. The small diagram shows the strategic range of herbs. Redrawn from J.R. Packham and E.V.J. Cohn (1990), *Arboricultural Journal*, **14**, 357–371 and reproduced by permission from A.B. Academic Publishers

seed, in some instances using mulches as a means of manipulation, both on their own and in conjunction with propagule containers placed at specific depths in relation to the substrate surface. A further experiment to assess the influence of two shade regimes (30% and 10% PAR) in combination with three nutrient treatments (+N, +P, +NP) is described elsewhere in Cohn (1994).

Compton litter experiment

In this experiment bulbs of *H. non-scripta* and well-established pot-grown plants of *L. galeobdolon* and *O. acetosella* were planted in a 9-year-old experimental oak (*Quercus* spp.) plantation (Hilton *et al.*, 1987) to which four litter treatments had been applied in a Latin square design (Figure 9.2). Seeds of *S. dioica* and foxglove (*Digitalis purpurea*) were then each sown at the rate of 100 seeds per square metre.

A	D	B	C
B	A	C	D
C	B	D	A
D	C	A	B

(a)

A: 2 cm deep Ercall oak leaf litter.

B: 4 cm deep Ercall oak leaf litter.

C: 3 cm deep seasoned woodchip.

D: Bare ground.

Ox	Hyac	Lam
Lam	Ox	Hyac
Hyac	Lam	Ox

(b)

Ox   *Oxalis acetosella*, wood sorrel
Hyac   *Hyacinthoides non-scripta*, bluebell
Lam   *Lamiastrum galeobdolon*, yellow archangel

Seeds of *Silene dioica*, red campion and *Digitalis purpurea*, foxglove, were sown at the rate of 100 m^{-2} over each plot.

Figure 9.2. Layout of Compton litter experiment: (a) Plan of experiment. 1 m wide buffer zone surrounds sixteen 2 × 2.3 m plots each with 4 litter treatments. (b) Plan of each 2 × 2.3 m plot

The early progress of this experiment is described in Cohn and Packham (1993). Subsequent events are even more important when attempting to assess the relative value of direct seedling as opposed to using pot-grown plants and of the influence of these particular mulches over the longer term. At the start of this experiment, the characteristics of this plot were predominantly those of a farmland or urban amenity plantation (Cohn and Packham, 1993), but after a relatively short period, both woodland edge and shade communities had become established. These were initially in the patterns imposed by the experimental design, but with successful propagation by self-sown seed in all species and extensive vegetative spread in *L. galeobdolon*, the communities are developing the mosaics more characteristic of ancient woodland (Packham *et al.*, 1992).

Establishment from seed in mulched and unmulched urban woodlands

In early spring 1990, this experimental work was extended into urban amenity plantations where mulches, when used, had been in place for several years. In the experiment described here, aspects of the early stages of woodland herb establishment from seed are examined in three urban plantations in which both a woodchip mulch and canopy closure alone have been reasonably effective in suppressing weeds.

SITES, MATERIALS AND METHODS

Sites

All three plantations were established on disturbed areas in Wolverhampton, and were planted between 1981 and 1985 with a variety of mainly broadleaved trees. At the time of planting a woodchip mulch was laid to a depth of 15 cm at the John Roberts Site (JR) and 10 cm at Phoenix Park (PH); the ground was left bare at Dunstall Hill (DU). The John Roberts plantation was established on brick rubble and Phoenix Park is a top-soiled former land-fill site. Dunstall Hill is the least disturbed of the three sites. There has been little or no further maintenance at any of the sites. The experiment was started in 1990, several years after canopy closure; there was little existing vegetation and none of the herbs to be introduced was present.

Experimental design

Four seed application treatments were involved:

- Treatment A: Unstratified seed sown in early spring 1990.
- Treatment B: Stratified seed (4 weeks at 5°C) sown in early spring 1990.
- Treatment C: Unstratified seed sown in autumn 1990.
- Treatment D: Stratified seed (as for B) sown in spring 1991.

All four seed application treatments were compared at John Roberts (JR) and Phoenix Park (PH), and the first three at the smallest site, Dunstall Hill (DU). Plots were arranged in a Latin square design (Figure 9.3) at John Roberts (16 plots) and Dunstall Hill (9 plots) to allow examination of site variability due, for example, to light gradients. At the irregular-shaped Phoenix Park site (16 plots), all four treatments were replicated four times, and allocated to plots in a randomised design.

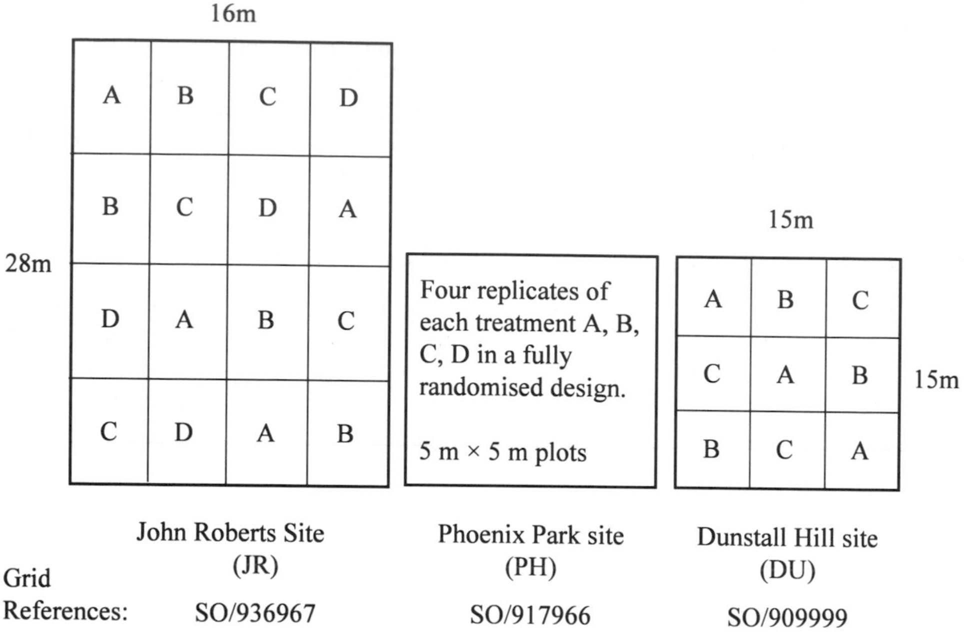

16m

A	B	C	D
B	C	D	A
D	A	B	C
C	D	A	B

28m

Four replicates of each treatment A, B, C, D in a fully randomised design.

5 m × 5 m plots

15m

A	B	C
C	A	B
B	C	A

15m

	John Roberts Site (JR)	Phoenix Park site (PH)	Dunstall Hill site (DU)
Grid References:	SO/936967	SO/917966	SO/909999

A: Unstratified seed sown in early spring 1990.
B: Stratified seed (4 weeks at 5°C) sown in early spring 1990.
C: Unstratified seed sown in autumn 1990.
D: Stratified seed (as for B) sown in spring 1991.

Figure 9.3. Plot layouts and treatments

Sowing treatments

A standard mix of seeds was used for each treatment (Table 9.2). Its composition was designed to deliver 100 seeds of each forb species per square metre, and graminoids at the rate of 25 seeds per square metre. Species were selected on the basis of their presence in local woodland herb communities and further depended on seed availability. They were also chosen to include a range of growth forms and ecological requirements. Apart from three of the grasses, false wood brome (*Brachypodium sylvaticum*), wood brome (*Bromus ramosus*), wood millet (*Milium effusum*), and pendulous sedge (*Carex pendula*), seed was obtained from commercial sources to assess its suitability.

Monitoring

Each plot was examined monthly through the spring and summer of 1990, and through the spring of 1991. Detailed recording of the central 3 m × 3 m of each plot was carried out in May/June 1990, June 1991 and May/June 1992. The number of seedlings of each species present, including sown herbs, natural colonisers and tree seedlings, was counted in each of the nine m² sub-quadrats. In addition, individual estimates of the area covered by all species present were made, including canopy trees.

Table 9.2. Composition of seed mix

Forbs		
Ct	*Campanula trachelium*	c
Dp	*Digitalis purpurea*	c
Fu	*Filipendula ulmaria*	c
Gu	*Geum urbanum*	c
Hns	*Hyacinthoides non-scripta*	c
Pv	*Primula vulgaris*	c
Sd	*Silene dioica*	c
Ss	*Stachys sylvatica*	c
Vr	*Viola riviniana*	c

Graminoids		
Bs	*Brachypodium sylvaticum*	1
Br	*Bromus ramosus*	1
Cp	*Carex pendula*	1
Me	*Milium effusum*	1
Pn	*Poa nemoralis*	c

c: commercial source; 1: local source.

USE OF COMPUTER ANALYSES TO DETECT CHANGES OCCURRING IN MANIPULATED WOODLANDS

The first stage in analysing the results of the experiment described above was to use TWINSPAN, two-way indicator species analysis (Hill, 1979) to compare the overall characteristics of the vegetation, both natural colonisers and the introduced herbs, to examine year to year changes and to assess differences both between sites and between plots. The results of this analysis, which are outlined below, will be described in greater detail elsewhere. The responses of individual herb species were then examined by analysing density records.

Since the purpose of the analysis was the examination of both vegetation characteristics at any one time, as well as changes from year to year, plot samples from C and D plots in the first year were also included, even though they had not yet had any seeds sown in them. A total of 123 plot samples were therefore analysed, 41 from each year.

TWINSPAN analysis

The major groups from the TWINSPAN analysis are shown in Figure 9.4: this shows the number of plot samples assigned to each group, and the indicator species which were significant in making the division, numbered in order of their effectiveness, and with the pseudospecies level shown in brackets. Introduced species are marked with an asterisk. Table 9.3 shows the actual number of samples assigned to each TWINSPAN group, broken down by treatment plot as well as by year and site. The first division of the 123 plot samples assigns 88 samples, a little over two-thirds of the total, into Group 0 and 35 samples to Group 1. Table 9.3 indicates that the samples have been divided principally by site, rather than by year.

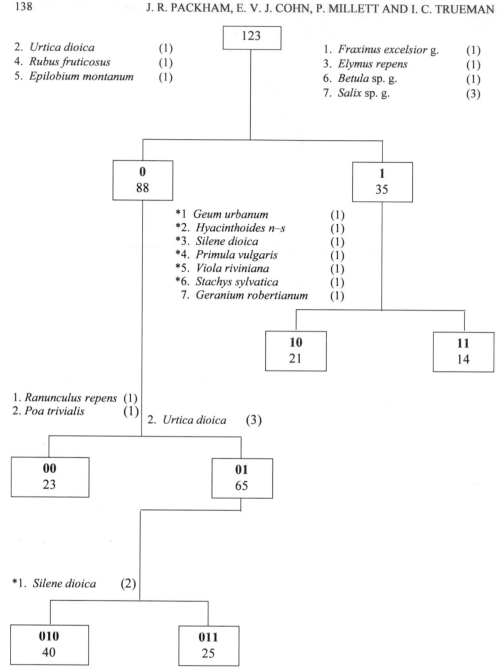

2. *Urtica dioica* (1)
4. *Rubus fruticosus* (1)
5. *Epilobium montanum* (1)

1. *Fraxinus excelsior* g. (1)
3. *Elymus repens* (1)
6. *Betula* sp. g. (1)
7. *Salix* sp. g. (3)

123

0
88

1
35

*1 *Geum urbanum* (1)
*2. *Hyacinthoides n–s* (1)
*3. *Silene dioica* (1)
*4. *Primula vulgaris* (1)
*5. *Viola riviniana* (1)
*6. *Stachys sylvatica* (1)
7. *Geranium robertianum* (1)

10
21

11
14

1. *Ranunculus repens* (1)
2. *Poa trivialis* (1)

2. *Urtica dioica* (3)

00
23

01
65

*1. *Silene dioica* (2)

010
40

011
25

Figure 9.4. TWINSPAN classification of the 123 (3m)² plot samples from the three sites in 1990, 1991 and 1992; all treatments, i.e. 41 plots each assessed 3 times. The upper figure in each box indicates the TWINSPAN group, the lower figure, the number of plot samples in the group. Indicator species are listed in order of importance with pseudospecies levels in brackets to indicate abundance; g. = ground layer, * introduced species

Table 9.3. TWINSPAN classification of plot samples from John Roberts (JR), Phoenix Park (PH) and Dunstall Hill (DU)

Year		TWINSPAN Group				
Site treatment		00	010	011	10	11
1990						
JR	A			3		1
	B			4		
	C			4		
	D			3		1
PH	A				2	2
	B			1		3
	C					4
	D			1		3
DU	A	2		1		
	B	2		1		
	C	3				
1991						
JR	A		4			
	B		4			
	C		2	2		
	D		2	2		
PH	A				4	
	B		1	1	2	
	C				4	
	D		1		3	
DU	A	2		1		
	B	2	1			
	C	3				
1992						
JR	A		4			
	B		4			
	C		4			
	D		3	1		
PH	A		1		3	
	B		3		1	
	C		3		1	
	D		3		1	
DU	A	3				
	B	3				
	C	3				

Chance and propagule availability play a significant role in determining the colonisers of urban habitats (Gilbert, 1990), so it should scarcely seem surprising that such an initial division should be made according to site. However, an examination of the indicator species for this division shows that all are either common perennial herbs or woody seedlings with parent trees present at all sites. Their identification by the computer program as indicator species might, therefore, be interpreted as being due to genuine differences in site conditions, rather than to chance. The indicators for Group 1 are generally less shade tolerant than those for Group 0; *U. dioica* and broad-leaved willowherb (*Epilobium montanum*) are also respectively more characteristic of nutrient- and base-rich situations. There is a progressive, though small, decrease in the mean percentage tree cover in each TWINSPAN Group from Group 00 (86%) to Group 11 (73%).

In contrast to the first division, the division of Group 1 divides the samples by year, all but two of the 1990 samples being assigned to Group 11 while all 21 of the 1991 and 1992 stands are placed in Group 10. The division here is clearly on the basis of the introduced herbs, with six of them appearing as indicators for Group 10.

As with the initial main division of the data, the division of the 88 stands in Group 0 separates plot samples from different sites rather than from different years. However, the subsequent division of Group 01 divides the plot samples by year, with the 1990 samples assigned to Group 011 and 1992 samples to Group 010 together with the majority of the 1991 plot samples. A single indicator, *Silene dioica* for the negative side, is sufficient precisely to define the dichotomy; it is present in 100% of samples in Group 010 and at a higher level of abundance than in the 16% of samples in which it occurs in Group 011. However, 11 other introduced herbs are also strongly preferential to this negative group; species such as wood avens (*Geum urbanum*), *H. non-scripta* and primrose (*Primula vulgaris*) being present in over 90% of the samples. This division then, appears to reflect the progressive establishment of the introduced field layer community at John Roberts and Phoenix Park.

Group 00 consists entirely of Dunstall Hill plots from the three years; the remaining four are in Group 011 (three) and Group 010 (one). Further division of this group indicates that within-site variability is greater than variability due to either year or treatment, although as will be discussed later, analysis of the introduced herb data shows that there are clear differences due to treatment.

The analysis so far has tended to divide plot samples in a way which identifies the differences between sites and between years; site differences are most marked between the mulched sites and the unmulched one, and year differences are largely attributable to the developing introduced herb communities. Further divisions of these major TWINSPAN groups start to divide plot samples according to sowing treatments, but these can be more effectively examined using density data in the next section.

It is instructive to examine Table 9.3, which confirms that the initial divisions of the data separate sample plots into groups from the same site or year. The table also confirms that despite the unavoidable heterogeneity of field sites subject to public amenity use and abuse, the initial plots, before being influenced by the sowing treatments (1990), were fairly homogeneous within any one site. It can also be seen from Table 9.3 that, over the three years, there is a trend towards greater homogeneity of plots, both within and between sites. This was shown at the division of Groups 1 and 01 to be due to the introduced field layer herbs.

Analysis of density records

Following a dry spring in 1990, no seedlings of any species were found when all plots were formally recorded in May 1990, nor were any found during earlier and later inspections of the plots. In June 1991, after another fairly dry spring, all species had germinated in at least one site, except for *M. effusum*, although its subsequent appearance in flower at all sites in 1992 suggests that it may have been overlooked in 1991. At least one species of introduced herb had germinated in all 3 m × 3 m plots, and in 80% of their component 1 m² sub-quadrats. There was considerable variability in the recorded densities at each site for any one treatment. This was apparent both at the scale of 9 m² (i.e. between plots), and also at the scale of 1 m² (within plots). These localised differences in successful establishment reflect not just environmental factors such as light and soil characteristics, but also the high level of disturbance associated with heavy amenity use, as Francis *et al.* (1992b) have also observed.

Distinct differences were observed between the responses of different species and in the responses of each species between treatments and between sites. These were additional to variation between replicated plots. *Geum urbanum* was the most successful species, occurring in all 41 plots in 1991, but five other species also had frequencies higher than the most frequent of the non-introduced herbs, *U. dioica* (Table 9.4).

Table 9.4. Ten most frequent herbs in 41 3 m × 3 m plots in 1991

	% frequency
Geum urbanum	100
Silene dioica	90
Hyacinthoides non-scripta	88
Primula vulgaris	85
Stachys sylvatica	63
Viola riviniana	61
Acer pseudoplatanus (seedlings)	56
Urtica dioica	51
Galium aparine	39
Rubus fruticosus agg.	36

Densities of the nine most frequent of the introduced herbs are shown in Table 9.5.

Table 9.5. Mean density of each species in 3 m × 3 m sample plots

		Gu	Sd	Hns	Pv	Ss	Vr	Bs	Br	Me	Pn	Ct	Dp	Fu	Cp
1991															
JR	A	127	26	256	30	2	5						1	2	
	B	87	21	76	23	6	5						1		
	C	88	49	36	28	9	3	3	4		13	<1		1	
	D	58	16	43	12	8	2	2	1		7	<1			
PH	A	101	10	28	14	1	1						1	1	
	B	233	25	35	39	8	10	1	1		3		1		<1
	C	125	11	40	21	6	5	8	3		4		1	1	
	D	124	14	19	1	<1	1	5	1				<1		
DU	A	11	<1		1							3			
	B	161	37	208	60	5	5					2			
	C	79	12	12	9	1		14							
1992															
JR	A	75	27	252	23	5	1					<1	1		1
	B	71	32	65	21	3	6		3	1		3	3	1	<1
	C	57	32	51	12	4	1	3	5	27			3		
	D	30	7	44	4	5	1	2	2	6					<1
PH	A	95	19	22	19	3	2						<1		
	B	125	76	11	14	5	3						<1		
	C	66	15	35	19	2	2	2	6	8					
	D	76	15	8	3	3		5	6	2					
DU	A	5				<1									
	B	206	131	205	63	2	5					2			
	C	60	15	60	4			12	1	2		2			

Species abbreviations as given in Table 9.2.

Although there are seemingly large differences between sites and between the four treatments at each site, these were not significant at JR and PH though they were at DU, the only site which did not have a woodchip mulch. Here, the A plots had a significantly lower number of seedlings of the four most frequent introduced herbs (p <0.05), and for *S. dioica*, *H. non-scripta* and *P. vulgaris*, the differences were also significant between all three treatments (A < C < B).

While treatment A was least successful at DU and PH, at JR it was the most successful overall, although the effect varied between species. Treatment B was the most successful at PH and DU and the most successful overall. Treatment D is comparable to the B treatment, but was rather less successful than the other three treatments for some forb species, most significantly *P. vulgaris* (p <0.01) and to a lesser extent *S. dioica*. It seems likely that harsh conditions shortly after sowing could account for this result.

There is little difference between the overall total density of all introduced herbs from 1991 to 1992 in the 3 m × 3 m plots. The density of some species has increased, though there is a slight drop overall (from 234 to 212 per 9 m^2). However, this drop is not significant, and was more than compensated for by a highly significant increase (p <<0.01) in the percentage cover of the introduced herbs in total, particularly *G. urbanum*, *S. dioica*, *P. vulgaris*, *S. sylvatica*, *B. sylvaticum*, *B. ramosus* and *M. effusum*.

While some general conclusions can be drawn from the broad trends shown in the establishment of the introduced herbs, more can be gained from an examination of the responses of the individual species to the four sowing treatments.

Geum urbanum

G. urbanum, one of the most successful species in this experiment, achieved its highest density in 1991, the year after sowing, in the B and C treatment plots at PH and DU. Its densities were generally lower at JR, the least disturbed site, and the one which is probably most nutrient and base rich, as indicated by the extensive cover of *U. dioica* (which was seen to increase markedly during the 3 years of the study). In 1990, this site had little by way of field layer apart from a few individuals of *U. dioica*. At this site, brick rubble lies beneath a thick layer of woodchip. It seems likely that the *U. dioica* is now responding to the phosphate of the brick rubble and that nitrogen, the one nutrient often lacking in brick rubble (Gilbert, 1989), is being supplied by the many alders present. The decline in the numbers of *G. urbanum* at this site from 1991 to 1992 may reflect both self-thinning and competition from *U. dioica* and *S. dioica*.

Silene dioica

Although not as fast-growing as *U. dioica*, *S. dioica* has a higher RGR (relative growth rate) than *G. urbanum* (Grime *et al.*, 1988) and is better able to exploit the high nutrient levels and thus compete with *U. dioica*. *S. dioica* was generally slower to establish at all sites than *G. urbanum* in 1991, but by 1992 its cover, if not its density, was approaching and at times exceeding that of *G. urbanum*. Both species have high germination rates, so it seems likely that the slower establishment of *Silene* reflects its greater drought sensitivity (Baker, 1947).

Hyacinthoides non-scripta

H. non-scripta seeds normally require chilling before they will germinate (Blackman and Rutter, 1954), so it is unlikely that the treatment A seeds will have germinated in 1991. It might be expected that treatments A and C seeds might have been more successful, though slower to germinate, than the treatments B and D seeds. This is because in separate germination experiments, naturally induced germination cues were found to be more successful than laboratory ones for those species with an obligatory chilling requirement. However, much depends on the conditions prevailing in the period following germination, and in this experiment weather conditions were particularly severe in both spring 1990 and 1991. The University of Wolverhampton weather station provided evidence of harsh dry conditions inimical to the seedlings of woodland herbs adapted to the shade and humidity of mature woodland.

That treatment B *H. non-scripta* seed, which had received their germination cues prior to their exposure to the 1990 dry spring conditions, were able to survive until the following spring suggests considerable resilience, and the ability to redirect their large seed reserve to produce a viable bulb for the following year with minimum photosynthetic activity.

The failure of both *H. non-scripta* and other seeds in the DU treatment A plots suggests that the losses due to predation and deterioration associated with a prolonged period in the soil, were exacerbated by the effects of drought in spring 1991. This would have been experienced more acutely by seedlings at DU, all of which lacked the protection of leaf litter or mulch.

Primula vulgaris and Viola riviniana

P. vulgaris and *V. riviniana*, both slow- and low-growing species without the advantages conferred by underground storage organs, are perhaps two of the species most vulnerable to the uncertain conditions of urban woodlands. They have similar habitat requirements, *P. vulgaris* being characteristic of slightly more moist conditions (Sinker *et al.*, 1985). It might, therefore, have been expected that their responses to sites and treatments would have been similar. However, *P. vulgaris* performed significantly better than *V. riviniana* with 5 to 6 times as many individuals becoming established under each treatment. Furthermore, *P. vulgaris* continued to flourish into the second year of establishment with a few individuals flowering, and many more approaching adult size. *P. vulgaris* density did not decrease significantly between 1991 and 1992. This was in contrast to *V. riviniana*, whose already low numbers declined further, with little growth in the few which did survive.

Unlike the forbs, the grasses got off to a somewhat slow start, with none appearing in the treatment A plots, and few in the treatment B plots. None of the grasses has a persistent seed bank and *M. effusum* is reported to be slow to germinate initially (Grime *et al.*, 1988). However, all the grasses, and *M. effusum* in particular, flowered profusely in 1992, so it might be expected that these species will continue to increase.

The small seeds of nettle-leaved bellflower (*Campanula trachelium*) and *D. purpurea* provide insufficient reserves for them to grow up quickly towards the light. They are consequently less able than the other herbs to take advantage of the more humid environment beneath a woodchip mulch. Neither species has done well; even at DU, where there is no mulch, they appear to have failed to establish in any numbers, possibly due to the dry springs of 1990 and 1991.

As might be expected in dry conditions, neither meadowsweet (*Filipendula ulmaria*) nor *C. pendula*, both species associated with damp conditions, have done well. Where *C. pendula* has established it might be expected to spread.

DISCUSSION

The influence of mulching

The presence of a woodchip mulch at two of the sites, JR and PH, appeared to mitigate some of the harsher features of the urban environment, where substrates are frequently dry and lacking in organic matter (Gilbert, 1989). At the third site, DU, without a mulch or even much leaf litter, there was little to protect seedlings from the severe fluctuations in soil and surface moisture likely to occur in a small site adjoining open space. Differences between sowing treatments were more marked at this site than the others, although once established, most species survived as well here as at the other two sites. Despite the unusually harsh spring conditions in the first two seasons of this experiment, all 14 species were able to establish in at least two sites and two treatments, and eight species flowered and set seed.

Time of sowing

While a consideration of the natural reproductive cycle of woodland herbs (with seed set and drying on the plant in the summer, followed by warm moist incubation in the soil until winter chilling and finally germination induced by warmer temperatures and longer day length) would suggest that autumn would be the best time to sow seed, the results of the C plots in this experiment suggest that there is no clear advantage in autumn sowing. There may be advantages in sowing seed at different times, not just to increase the chance of at least some germination under favourable conditions, but also to ensure that there is opportunity for the full range of genetic variability with respect to germination to be expressed. In this way, the population as a whole will possess greater resilience to the spatially and temporally variable environment in urban amenity woodlands.

Seed or transplants?

It is this greater potential for variability and the resilience this confers which make seed the preferred means to enrich the field layer of secondary woodlands. However, some species do establish best from transplants and these should be planted in early spring. These include shade tolerant taxa such as *O. acetosella* and *L. galeobdolon*, both of which form clonal patches but whose seeds are very difficult to store (Packham, 1978; 1983). Cost is clearly another factor; despite great variability in reproductive capacity (Salisbury, 1942), it is still cheaper to allow the seed of many species to take 'pot-luck' among the spatial mosaic of favourable and unfavourable areas in urban woodlands than it is to introduce nursery reared plants. Plants in the juvenile phase may be subject to different selection pressures from adult plants, or may respond in different ways to the same selection pressure (Grime, 1979). It cannot be assumed, therefore, that an individual will be able to replace itself in a situation where it has been successfully planted as a mature plant.

Whether seed or plants are the chosen material, problems can arise if it is of unsuitable provenance. *V. riviniana* in the experiment described here failed to thrive even in conditions where it might have been expected to, alongside some of its common associates. The most likely explanation for this is that the commercially supplied seed is from a population from an unshaded habitat. Such populations are genetically and physiologically different from those from woodland habitats (Valentine, 1941). Fu (1993) reports a failure of germination, or at least of development into detectable plants, under field conditions in similar commercial seed of this species.

In addition to ecological and economic considerations, there is a third advantage in using seed rather than plants. Much basic information can be gained by observing the ability of native species to establish and thrive in a range of situations. This information can then be used in the reconstruction of more authentic woodland communities. The concept of habitat reconstruction as an heuristic activity, an opportunity to gain knowledge as well as to deploy it, is a theme developed by Harper (1987). Bradshaw (1987) considered the true test to be the ability to restore 'in proper form or amount, all the components of the ecosystem we infer to be crucial, and then find that we have recreated an ecosystem that is indistinguishable in both structure and function from the original ecosystem, or the ecosystem that served as our model'. Habitat creation with this ambitious goal is a contentious issue in this country, particularly with regard to the reconstruction of semi-natural woodlands, and what has always been difficult may now be rendered even more so by climatic change. Warmer springs could upset the delicate balance between the growth phases of vernal geophytes and competitive perennials such as *P. aquilinum, Holcus mollis* and rough meadow-grass (*Poa trivialis*). Massive DNA content and large cell size enable woodland geophytes such as *H. non-scripta* and wood anemone (*Anemone nemorosa*) to grow rapidly in the cold temperatures of early spring by the expansion of large cells formed during the previous summer (Grime *et al.*, 1988). In contrast, the growth of their potential competitors is constrained by their greater frost sensitivity (e.g. *P. aquilinum*) or their dependence on warmer temperatures for growth.

CONCLUSIONS

There is now a much firmer basis upon which planners and managers of farm and urban woodlands can build. Where the intention is to create communities approaching those of native woodlands, Rodwell and Patterson (1994) provide an excellent basis for the selection of tree and shrub species suitable for the geology and region concerned. Former arable land and urban sites frequently resemble each other in having high nutrient status and large seed banks of aggressive ruderal species, which make the initial use of herbicide treatments, followed by mulching, highly desirable. Even so, species of high relative growth rate (RGR) and those encouraged by high P and N levels, of which *U. dioica* is a notorious example, can only be controlled in the long term by shading; the high mineral nutrient content of many soils severely limits the type of field layer which can be established.

The stage at which field layer species can be successfully introduced will vary considerably with the soils, trees and shrubs concerned but with pedunculate oak (*Quercus robur*) planted at 1.0-1.5 m centres a suitable time would be the spring of the sixth season. Thought should be given at an early stage to the desired mosaic pattern. *O. acetosella* and *L. galeobdolon* may occur scattered amidst other species in nature; they can also occur in quite large patches, a situation which can be encouraged by creating areas of heavy shade. If starting from scratch it may be worth scraping off the nutrient and seed-rich surface soil to provide an opportunity for many of the less competitive herbs and dwarf shrubs. Where soil is dumped, the placing of sandy loam material in some areas where *H. non-scripta* and *A. nemorosa* will flourish, and of damp heavy soils in others where *H. non-scripta* will die, but species such as tufted hair-grass (*Deschampsia cespitosa*), *Filipendula ulmaria* and valerian (*Valeriana dioica*) establish well, will itself strongly influence the vegetational pattern. In the long term the most sensible practice is to reduce subsequent management by creating an environment which favours the desired community as much as possible.

Acknowledgements

The experiments described in this chapter were completed by one of us (E.V.J.C.) during the tenure of a research studentship awarded by the University of Wolverhampton. We are grateful to Wolverhampton Metropolitan Borough Council for their support and the provision of field sites and materials, to Professor J. P. Grime for helpful discussion in the early stages of this work and to Dr R. Binns for statistical advice.

REFERENCES

Ashby, W.C. (1987). Forests. In: Jordan, W.R. III, Gilpin, M.E. and Aber, J.D. (eds). *Restoration Ecology: a Synthetic Approach to Ecological Research*. Cambridge University Press, Cambridge, 89–108.

Baker, H.G. (1947). *Melandrium dioicum* (L. emend.) Coss. and Germ. Biological Flora of the British Isles. *Journal of Ecology*, **35**, 283–292.

Blackman, G.E. and Rutter, A.J. (1954). *Endymion nonscriptus* (L.) Garcke. Biological flora of the British Isles. *Journal of Ecology*, **42**, 629–638.

Bradshaw, A.D. (1987). Restoration: an acid test for ecology. In: Jordan, W.R. III, Gilpin, M.E. and Aber, J.D. (eds). *Restoration Ecology: a Synthetic Approach to Ecological Research*. Cambridge University Press, Cambridge, 23–30.

Brown, R.J. (1989). Wild flower seed mixtures: supply and demand in the horticultural industry. In: Buckley, G.P. (ed.). *Biological Habitat Reconstruction*. Belhaven Press, London, 201–214.

Buckley, G.P. and Knight, D.G. (1989). The feasibility of habitat reconstruction. In: Buckley, G.P. (ed.). *Biological Habitat Reconstruction*. Belhaven Press, London, 171–188.

Cohn, E.V.J. (1994). The manipulation, introduction and ecology of field layer communities in broadleaved woodlands. PhD thesis, University of Wolverhampton.

Cohn, E.V.J. and Packham, J.R. (1993). The introduction and manipulation of woodland field layers: seeds, plants, timing and economics. *Arboricultural Journal*, **17**, 69–83.

Down, C.G. and Morton, A.J. (1989). A case study of whole woodland transplanting. In: Buckley, G.P. (ed.) *Biological Habitat Reconstruction*. Belhaven Press, London, 251–257.

Ellenberg, H., Weber, H. E., Duell, R., Wirth, V., Werner, W., Paulissen, D. (1991). Zeigerwerte von Pflanzen in Mitteleuropa. *Scripta Geobotanica*, **18**, Verlag Erich Goltze KG, Göttingen.

Francis, J.L., Morton, A.J. and Boorman, L.A. (1992a). The establishment of ground flora species in recently planted woodland. *Aspects of Applied Biology*, **29**, 171–178.

Francis, J.L., Morton, A.J. and Boorman, L.A. (1992b). The introduction of woodland herbaceous species into secondary woodland. In: Packham, J.R. and Harding, D.J.L. (eds). *Woodland Establishment, Maintenance and Assessment*. Wolverhampton Woodland Research Group, Wolverhampton, 15–26.

Fu, D.L. (1993). A physiological and ecological study of the feasibility of establishing field layer vegetation in urban woodlands. PhD thesis, Wye College, University of London.

Gilbert, O.L. (1989). *The Ecology of Urban Habitats*. Chapman and Hall, London.

Grabham, P.W. and Packham, J.R. (1983). A comparative study of the bluebell *Hyacinthoides nonscripta* (L.) Chouard in two different woodland situations in the West Midlands, England. *Biological Conservation*, **26**, 105–126.

Grime, J.P. (1979). *Plant Strategies and Vegetation Processes*. Wiley, Chichester.

Grime, J.P., Hodgson, J.G. and Hunt, R. (1988). *Comparative Plant Ecology: a Functional Approach to Common British Species*. Unwin Hyman, London.

Grubb, P.J. (1977). The maintenance of species-richness in plant communities: the importance of the regeneration niche. *Biological Reviews*, **52**, 107–145.

Harper, J.L. (1987). The heuristic value of ecological restoration. In: Jordan, W.R. III, Gilpin, M.E. and Aber, J.D. (eds). *Restoration Ecology: a Synthetic Approach to Ecological Research*. Cambridge University Press, Cambridge, 35–46.

Hill, M.O. (1979). TWINSPAN – a FORTRAN program for arranging multivariate data in an ordered two-way table by classification of the individuals and attributes. Section of Ecology and Systematics, Cornell University, Ithaca, New York.

Hilton, G.M., Packham, J.R. and Willis, A.J. (1987). Effects of experimental defoliation on a population of pedunculate oak (*Quercus robur* L.). *New Phytologist*, **107**, 603–612.

Hodgson, J.C. (1989). Selecting and managing plant materials used in habitat construction. In: Buckley, G.P. (ed.). *Biological Habitat Reconstruction*. Belhaven Press, London, 45–67.

Knight, G.H. (1964). Some factors affecting the distribution of *Endymion non-scriptus* (L.) Garke in Warwickshire woods. *Journal of Ecology*, **52**, 405–421.

Packham, J.R. (1978). *Oxalis acetosella* L. Biological Flora of the British Isles. *Journal of Ecology*, **66**, 669–693.

Packham, J.R. (1983). *Lamiastrum galeobdolon* (L.) Ehrend. and Polatschek. Biological Flora of the British Isles. *Journal of Ecology,* **71**, 975–997.

Packham, J.R. and Cohn, E.V.J. (1990). Ecology of the woodland field layer. *Arboricultural Journal,* **14**, 357–371.

Packham, J.R., Harding, D.J.L., Hilton, G.M. and Stuttard, R.A.S. (1992). *Functional Ecology of Woodlands and Forests.* Chapman and Hall, London.

Packham, J.R. and Willis, A.J. (1976). Aspects of the ecological amplitude of two woodland herbs, *Oxalis acetosella* L. and *Galeobdolon luteum* Huds. *Journal of Ecology*, **64**, 485–510.

Packham, J.R. and Willis, A.J. (1982). The influence of shading and of soil type on the growth of *Galeobdolon luteum. Journal of Ecology*, **70**, 491–512.

Peterken, G.R. (1981). *Woodland Conservation and Management.* Chapman and Hall, London.

Rodwell, J.S. (ed.) (1991) *British Plant Communities*, vol. 1: *Woodlands and Scrub.* Cambridge University Press, Cambridge.

Rodwell, J.S. and Patterson, G. (1994). *Creating New Native Woodlands.* Forestry Commission Bulletin 112. HMSO, London.

Ruff, A.R. (1979). *Holland and the Dutch Ecological Landscapes.* Deanwater Press, Manchester.

Salisbury, E.J. (1916). The emergence of aerial organs in woodland plants. *Journal of Ecology,* **4**, 121–128.

Salisbury, E.J. (1942). *The Reproductive Capacity of Flowering Plants.* Bell, London.

Seed, R.F. (1993). Use of seed mixtures in modifying field layer vegetation in plantations in Milton Keynes. MSc thesis, University of Wolverhampton.

Sinker, C.A., Packham, J.R., Trueman, I.C., Oswald, P.H., Perring, F.H. and Prestwood, W.V. (1985). *Ecological Flora of the Shropshire Region.* Shropshire Trust for Nature Conservation.

Sydes, C. and Grime, J.P. (1981). Effects of tree leaf litter on herbaceous vegetation in deciduous woodland. II. An experimental investigation. *Journal of Ecology*, **69**, 249–262.

Trueman, I.C., Sinker, C.A. and Packham, J.R. (1985). Habitats and plant communities. In: Sinker, C.A. Packham, J.R., Trueman, I.C., Oswald, P.H., Perring, F.H. and Prestwood, W.V. (eds). *An Ecological Flora of the Shropshire Region.* STNC, Shrewsbury.

Valentine, D.H. (1941). Variation in *Viola riviniana* Rchb. *New Phytologist*, **40**, 189–209.

10 Invertebrate Conservation and New Woodland in Britain

R. S. KEY

INTRODUCTION

Almost any group of newly planted trees will, fairly rapidly, attract quite a large and sometimes diverse invertebrate fauna. The planting of trees may contribute significantly to the production of biomass of invertebrates to provide food for vertebrates, especially birds. However, whether this contributes significantly to the real conservation needs of the more vulnerable woodland or other invertebrates has been questioned (e.g. Kirby, 1992). The purpose of this chapter is to highlight the contribution that the creation of new woodland can have for invertebrate conservation and how this can be maximised, and also to indicate its limitations so that expectations are not set too high.

Woodland invertebrates and their habitats

Old woodland is the richest habitat for terrestrial invertebrates, and almost all groups of invertebrates have large numbers of woodland species. Even among exclusively aquatic groups, there are species specialised for living in pools or streams only in woodland. The diversity of niches provided by its great plant architectural diversity and the relative stability of conditions produced in a climax but very dynamic ecosystem has resulted in a naturally very diverse fauna. There are particular assemblages of species associated with the pioneer, intermediate and climax phases in the woodland succession (Warren and Key, 1991). In lowland temperate mature woodland, especially that which includes plenty of open space and mature trees, the number of species of invertebrates present may run to several thousands (Elton, 1966). Elton summarised the most important microhabitats in temperate woodland (Table 10.1).

A number of features common to many invertebrates makes historical continuity of habitat conditions vitally important. Many species are extremely specialised in their habitat requirements and they may therefore be dependent on very small, sometimes infrequent (either or both in space or time), and easily overlooked, features in their habitat. The adults and larvae of many species of insect frequently have very different habitat requirements, sometimes occurring in totally dissimilar habitats from each other that therefore need to be within the dispersal range of each life stage. This makes a dynamic and integrated mosaic of habitat conditions, in particular of seral stages, very important. However, many species seem to have limited powers of dispersal or colonisation (Hammond, 1974), either through actual limitations in mobility, such as some

The Ecology of Woodland Creation. Edited by Richard Ferris-Kaan.

Table 10.1. Important invertebrate habitat in woodland: features particularly important to open space species italicised; features particularly important to old forest species emboldened

Major feature	Habitat	Microhabitat		Use by invertebrates
Woody plants trees and shrubs	Foliage	Canopy		Mainly as foodplants
		Understorey		Mainly as foodplants
		At woodland/ridge edge		*Foodplants/perches etc.*
	Flowers	*Sallow, hawthorn, sloe*		*Nectar sources*
	Bark	Surface	Surface	Perches and predators
			Rugosities	**Niches for cover**
		Damage	**Loose bark**	**Saproxylics, cover**
			Sap runs	**Specialists**
	Living wood			A very few woodborers
	Damaged or decaying wood on living trees	**Rot holes**		**Saproxylics, cover**
		Heart rot		**Saproxylics**
		Cavities		**Saproxylics, cover**
	Water-filled rot holes			**Specialist fauna**
	Roots	Living		Mainly gall causers
		Decaying		**Saproxylics**
Herbs	Foliage	*In sun*		*Foodplants, perches etc.*
		In shade		Foodplants
	Flowers/nectar sources	*Especially composites and umbellifers*		*Nectar sources*
	Roots/storage organs	Mainly bulbous species		Foodplants
Bryophytes				Cover
Fungi	Soil fungi and mycorrhizae	Mycelium		Specialist mycophages
		Fruiting bodies		Specialist mycophages
	Dead wood fungi	**Mycelium**		**Specialist saproxylics**
		Fruiting bodies	**Bracket**	**Specialist saproxylics**
			Gill	**Specialist saproxylics**
Plant litter	Leaf litter			Cover, saprophages
	Fallen dead wood	Accumulations of twigs		Specialist saproxylics
		Branches		Saproxylics
		Tree boles		**Saproxylics**
		Wood mould		**Specialist saproxylics** and soil fauna
Soil	Soil surface	*Bare soil in sun*		*Solitary bee and wasp nests* *Basking – thermophilic species*
	Soil body	Woodland soil in shade		Specialist soil fauna
	Soil water	Seepages and flushes		Specialist 'squidge' fauna Different in *sun* and shade
Water	Standing water	Ponds and lakes		Different faunas in *sun* and shade
		Puddles (mainly in sun)		*Temporary pool fauna on clay*
	Running water	Rivers and streams in shade		A few specialist aquatics

hygrophilous woodland molluscs, or from an apparent 'reluctance' to set off for pastures new. The life cycle of most invertebrates is strictly annual, with no 'resting stage' through which a species might survive periods of adversity. This makes continuity (though not necessarily stability) of conditions particularly important.

The above features in combination make many species of woodland invertebrate very vulnerable to changes in their habitat, such that they may easily become extinct but are often very slow to recolonise or colonise new habitats. Habitat with a long temporal continuity of particular conditions is therefore very important and this has been shown especially to be the case for species associated with woodland (e.g. Harding and Rose, 1986). Even for pioneer communities, a continuity of regular dynamic change is necessary for these communities to persist in an area (e.g. Warren, 1987). Many woodland invertebrates will, therefore, be very slow to colonise newly planted woodland.

Of course, the above limitations do not apply to all species of woodland invertebrates. Many are highly mobile and will rapidly colonise new plantations almost as soon as the right conditions become available. These species are the commoner, less fastidious ones, least in need of conservation measures to ensure their survival. The provision of more habitat for these species to colonise is, nevertheless, a valuable exercise in its own right. Many such species add much to human enjoyment of woodland and contribute to the development of a functioning woodland ecosystem, including providing food for other organisms. However, without careful consideration of the specific needs of the more fastidious woodland invertebrates, those species most in need of conservation are likely to benefit least from the provision of new woodland habitat.

Colonisation of new woodland by invertebrates

There seem to have been relatively few studies of the colonisation rates into new woodland. Casual observations on woods that have developed on non-wooded land several decades ago indicate that they may now hold very rich faunas, including some scarce species. The extensive birch (*Betula pendula*) woodland of Holme Fen National Nature Reserve in Cambridgeshire developed on farmland within the last 120 years but now supports an exceptional fauna including nationally scarce species (Kirby, 1987). However, such woods developed at a time when the surrounding countryside was much more diverse and richer in species to colonise them. Whether such a fauna will develop over the same timescale in woodlands planted from scratch now is uncertain.

DESIGNING AND MANAGING HABITAT FOR INVERTEBRATES IN NEW WOODLAND

Many features can be incorporated into new woodland that can enhance its conservation value for invertebrates and the likelihood of colonisation by woodland species.

Location of new woodland

The location obviously may be something over which we have little or no control, except whether to plant or not (see Harmer and Kerr, Chapter 8): this is an important initial question. However, the location of new woodland, particularly in relation to other habitat, other woodland and to possible corridors for colonisation (usually hedgerows, watercourses and other linear features such as verges and railway lines) could make considerable difference to the colonisation rate and subsequent value for invertebrates (Spellerberg and Gaywood, 1993; Dawson, 1994).

It is very rare that the planting of new woodland directly onto existing semi-natural habitat is likely to be to the benefit of invertebrate conservation. More likely, a fauna of existing value may be eliminated and replaced with a fauna of lesser interest. The invertebrates of habitats such as heathland, grassland and any type of wetland are under such threat that it is inappropriate for existing habitat to be considered for conversion to woodland. Even the planting of individual trees on such habitats may threaten their invertebrate fauna, usually through shading of basking and nesting sites. It is of far greater benefit to restore appropriate management on even the most degraded examples of such habitat than to convert it to woodland. Woodland creation and tree planting may, surprisingly, also be detrimental to existing invertebrate faunas in apparently unlikely situations such as industrially derelict sites, spoil and old mineral workings; types of land where woodland or other greenspace creation is often targeted. Such 'wasteland', especially if on very infertile soil, may have an almost arrested plant succession, with much bare ground and ruderal vegetation. Pioneer communities of invertebrates, sometimes with a strong heathland or calcareous grassland element, have often colonised such sites while becoming extinct in nearby more 'natural' habitats, simply through natural plant succession (Lonsdale, 1991). Old clay and sand workings, mine, furnace or ash spoil and even ruins of old buildings and areas of old broken concrete, may be refuges for species otherwise long gone from an area. Before considering 'restoration' of such long-established sites, it is recommended that a survey of the invertebrate fauna be carried out.

The incorporation of existing habitat either within, or adjacent to, new woodland is another matter entirely. Many species of invertebrates typically inhabit interfaces or successions between habitat types (Kirby, 1992), and many thermophilic species will benefit from increased warmth and shelter from the wind provided by adjacent tree cover. The increased structural diversity created by the woodland and its edges will also benefit many species that need perches, territorial markers, etc. Existing habitat may also form a source of colonisation for certain elements of the woodland fauna. For example, the edges of rides and glades in new woodland would be more likely to be colonised by incorporating existing scrub into the woodland. In particular, the incorporation of existing woodland, hedgerows and individual trees is likely to contribute significantly to new woodland fauna, especially if these features can be maintained as internal edges within the new woodland, on the margins of rides and glades.

There are also sometimes dangers to existing faunas to guard against when incorporating blocks of existing habitat into new woodland, or situating new woodland adjacent to them. The most important to guard against is the possibility of shading out or scrubbing over of open sunny habitats as the planted trees mature and start to regenerate. The drying out of existing damp and wet habitat through the increased evapo-transpiration caused by the new trees may also be a danger. Unless the hydrology of such wet areas can be safeguarded, the appropriateness of woodland creation should be questioned.

All such incorporated habitats should be specifically managed as woodland open space and steps taken to ensure that trees on their wooded margins, especially on the southern, eastern and western margins, do not shade this open space. General advice on the management of open space in woodland is given by Kirby (1992) and Warren and Fuller

(1993) and should be followed in almost any new woodland. Tree planting should not isolate existing non-woodland habitat from other nearby examples of the same habitat by the creation of barriers to movement. Indeed, if corridors of similar habitat can be created along with the woodland, so much the better. Benefits of, and precautions to take into account in the incorporation of, existing habitat into new woodland are given in Table 10.2.

Table 10.2. Benefits and problems to invertebrate conservation of incorporating blocks of other habitat in new woodland and potential management solutions

Incorporated habitat	Benefits	Problems	Possible solution
Woodland	Source of colonisation of new woodland Enhanced viability of fauna of small woodland	Edge fauna may suffer	Maintain internal edges
Hedgerow and trees	Source of colonisation of new woodland More diverse tree age structure in new woodland Good nectar sources	Overtopping or shading of shrubs or older trees	Maintain as internal edges
Scrub	Source of colonisation of wood and ride edges Good nectar sources	Overtopping or shading of shrubs Rarely appreciated and often neglected Eventually becomes too mature and fills available space	Maintain as internal edges Education Manage on coppice style rotation
Grassland and heathland	Source of colonisation of rides and glades Increased shelter and warmth	Danger of shading small patches Tree/shrub regeneration into habitat More remote from graziers or difficult access for machines	Coppice edges, especially on south and east sides Manage specifically as open space Design ride system to allow easy access
'Waste' ground (including broken concrete, tarmac, rubble etc.)	Ruderal nectar plants Sunny bare ground for basking and burrowing Cover and shelter	Rarely appreciated Danger of shading small patches Tree/shrub regeneration into habitat	Education Coppice edges, especially on south and east sides Manage specifically as open space
Wetland	Increased shelter Less likelihood of pesticide or fertiliser input	Danger of shading small patches Increased evapotranspiration leading to drying out Nutrient input from leaf fall	Coppice edges, especially on south and east sides Is woodland planting appropriate here? As above

Overall management regime of new woodland

The fauna that develops and persists in new woodland will depend greatly on its subsequent management.

Coppice

Traditionally managed coppice supports a very rich fauna of invertebrates typical of open space in woodland and scrub margins, many of which have declined drastically in Britain with the widespread demise of coppicing over the last 50 years. Whether it is appropriate to introduce a coppicing regime to all or part of a block of new woodland, at least from the point of view of benefit to the invertebrate fauna, depends greatly on existing, currently coppiced woodland being in quite close proximity as a source of colonisation. Many coppice woodland invertebrates seem to be poor colonisers (e.g. Warren, 1987) and may not take advantage of the conditions so created unless already present nearby. Such management may, however, benefit many more generalist species, many of which will thrive in the open conditions provided in coppice panels or on the foliage on the coppice regrowth. The benefits and techniques of coppice management for invertebrates are described by Fuller and Warren (1990).

Pasture woodland

A rare and declining form of woodland management is wood-pasture, where trees are usually pollarded and animals grazed between them. This method of management has allowed pollarded trees to reach great age and therefore become of particular value to invertebrates associated with a long continuity of decaying timber (Harding and Rose, 1986). Some species are restricted to the very oldest trees. Invertebrates in this habitat seem to have particularly poor powers of colonisation (Warren and Key, 1991) and existing sites form important refuges for nationally and internationally threatened species. A particular problem in such sites is a skewness in tree population structure towards very old individuals, resulting from the long history of grazing preventing regeneration. A very significant contribution to the future viability of the invertebrate fauna of such sites could result from the creation of new wood-pasture adjacent to existing sites, with subsequent long-term management of the trees as pollards. It is very dubious whether the creation of new wood-pasture in isolation from existing sites with very mature trees would contribute significantly to invertebrate conservation.

High forest

The invertebrate fauna of high forest is rich in species, although it includes relatively few species that face conservation problems, presumably a result of the conversion of so much coppice to high forest in Britain in recent decades. A typical high forest fauna is likely to develop as the woodland matures, although the greatest contribution to woodland invertebrate conservation, particularly for the scarcer species, from new high forest may well be from its areas with no trees, that is, its ride and glade system.

Provision of open space in new woodland

The provision of open, sunny space within woodland is very important for most woodland invertebrates. Many typical woodland species are thermophilic nectar feeders as adults, even though their larval stages may require cool, moist or shady conditions. The creation of new woodland gives the opportunity to incorporate large glades and a wide system of interconnected rides in the best positions at the design stage. Not only will the invertebrates themselves benefit from a well laid out system of open space, but the more conspicuous elements of that fauna will be more likely to be enjoyed by human visitors to the woodland. Only in very wet conditions, where stable alder (*Alnus glutinosa*) or willow (*Salix* spp.) swamp or carr may be the end result of woodland creation and where shade and high humidity are of particular importance to the invertebrate species (A. Stubbs, personal communication), is the incorporation of open sunny space inappropriate.

Open space in new woodland can be designed to maximise the possibility of forming suntraps and of reducing shade while maximising shelter from wind. The incorporation of any south or east facing slopes into woodland open space will produce benefits for invertebrates, as will maximising the length of internal edge habitat between the woodland and its open space, particularly on northern (south-facing) edges. This can be increased by designing irregularity into the margin, e.g. by scalloping of ride edges as described by Carter and Anderson (1987) and Warren and Fuller (1993). A circumference ride to the woodland, separated from surrounding land by sturdy flowering hedge system, could also contribute significantly to this open space and edge fauna, as well as protecting the wood from pollutants such as spray drift.

Open space in new woodland is best managed as grassland, heathland or wetland depending on prevailing soil conditions, with maintenance of a varied physical structure of the vegetation and provision of abundant nectar plants of the appropriate species (see below). Rotational mowing and raking of vegetation all help in the provision of such a varied vegetation structure, and guidelines are given by Warren and Fuller (1993).

Open space in woodland will ideally include unvegetated ground open to the sun. Many species of invertebrates typical of woodland and other habitats use bare ground for basking, for the creation of nest burrows and for hunting over (Plate 10.1), for example solitary bees of the genus *Lasioglossum* and *Andrena* and the conspicuous bee fly (*Bombylius major*). Unsurfaced vehicle or pedestrian trackways with slight surface erosion often provide ideal habitat for such species, although that created by galloping horses is not as valuable, as the churned up soil does not provide habitat for burrowing species. Eroded areas such as this should be regarded as positive features, and access should not be prevented in order to allow them to revegetate, a practice that is often carried out for apparent conservation benefits. It may actually be worthwhile to introduce rotational scraping of surface vegetation in sunny areas to maintain such bare ground. Such bare ground is particularly valuable if sloping towards the sun. In ancient woodland, old wood banks often provide useful examples of this habitat and the creation of new woodbanks along the northern and western margins of rides and glades at the design stage could be considered. Guidance on the management of bare ground is given by Key and Gent (1993).

Plate 10.1. Bare ground in rides and glades in new woodland may be used by a wide variety of invertebrates, to bask on, burrow into and hunt over. On sandy ground, the digger wasp *(Cerceris arenaria)* may do all three (R.S. Key, English Nature)

Tree and shrub species

Some species of trees support much richer faunas of plant-feeding invertebrates than others (Kennedy and Southwood, 1984) and are therefore more valuable in producing a rich fauna. However, as many plant-feeding invertebrates are highly species-specific, restriction of planting only to the species supporting the largest number of invertebrates is not justified (Plate 10.2). It is best to plant a range of tree and shrub species natural to woodland in the area in which the new woodland is planted.

In general, native tree and shrub species support more species of invertebrate than introduced ones, although some species of southern beech (*Nothofagus spp.*) have attracted several oak (*Quercus* spp.) and beech (*Fagus sylvatica*) feeders (Welch and Greatorex-Davies, 1993). However some non-native trees, including conifers such as spruce (*Picea* spp.), larch (*Larix* spp.) and pine (*Pinus* spp.) (Scots pine (*Pinus sylvestris*) considered non-native outside the Caledonian Forest area of Scotland and possibly the East Anglian Breckland, where its status as a native species is enigmatic) also support many species of invertebrate. Most of these are likely to be recent colonists to Britain and whether attracting such species to our new planted woodland by including non-native species of tree really contributes to wildlife conservation is open to debate. It is reasonable to argue that targeting planting to benefit invertebrates associated with native tree species is more appropriate. The number of foliage-eating and sap-sucking species associated with both native and non-native trees is summarised in Table 10.3.

Table 10.3. Number of plant-feeding species of invertebrate associated with various native and introduced trees in Britain. Largely derived from C.E.J. Kennedy and T.R.E. Southwood, 1984, *Journal of Animal Ecology*, **53**, 455–478

Native or introduced	Broadleaved or conifer	Tree species	Number of species	Comment
Native	Broadleaved	Willows/sallows	450	Important nectar provider
		Native oaks	423	
		Birch	334	
		Hawthorn	209	Very important nectar provider
		Poplar/aspen	153	Important nectar provider
		Blackthorn	153	Important nectar provider
		Alder	141	
		Elm	124	
		Apple	118	
		Hazel	106	
		Beech	98	
		Ash	68	
		Rowan	58	
		Lime	57	Useful nectar provider
		Field maple	51	
		Hornbeam	51	
		Holly	10	
	Conifer	Scots pine	172	Native in Scotland (and the Breckland ?)
		Juniper	32	
		Yew	6	
Introduced species	Broadleaved	Southern beech	78[a]	Data from Welch and Greatorex-Davies, 1993
		Sycamore	43	
		Sweet chestnut	11	
		Horse chestnut	9	
		Walnut	7	
		Holme oak	5	
	Conifer	Spruce	70	
		Larch	38	

[a] Refers to Lepidoptera only.

If one of the objectives of planting the new woodland is to augment existing sites with old trees that may support deadwood feeding invertebrates, the above tabulated 'ranking' of tree species for foliage-eating invertebrates is not relevant. Similar numerical analysis of wood feeders on different trees has not been carried out and the state and type of wood decay, rather than species of tree, may be as or more important to the invertebrates. General observations have, however, indicated that native species, especially oak (*Quercus petraea* and *Q. robur*), beech (*Fagus sylvatica*), elm (*Ulmus* spp.), birch (*Betula* spp.), larger willow species (*Salix* spp.) and lime (*Tilia* spp.), are especially useful. Outside Scotland and the Breckland, where Scots pine (*Pinus sylvestris*) supports rich faunas of saproxylic species (Foster, 1986), conifers are of little value. Of the non-native broadleaved trees, horse chestnut (*Aesculus hippocastanum*) is perhaps the most useful in that it is, like *Ulmus* spp., quite prone to permanent sap runs that are habitat for some very specialised species that have declined with the advent of Dutch elm disease.

Nectar and pollen plants

Plants with easily accessible pollen and nectar are important to very many woodland invertebrates, and their inclusion in new woodland is likely both to enhance its amenity value and allow people actually to see the more conspicuous woodland and other invertebrates. Not only do many invertebrates rely on nectar and pollen as sources of energy and protein while adults, many predatory species such as the soldier beetles and some spiders specialise in feeding on flower-visiting insects (Kirby, 1992). Consideration of the flowering season of nectar plants included in a planting scheme can ensure that the entire spring to autumn period is covered, preferably aiming for an early summer peak in 'floweriness' when a majority of woodland insects are adult. Open-structured flowers such as umbellifers, composites and members of the *Rosaceae* are more likely to attract a diversity of species than tubular-flowered species that require specialist mouthparts. Both flowering shrubby species and herbaceous species are best included.

Particularly useful early-flowering nectar plants are the willows and sallows (*Salix* spp.), blackthorn (*Prunus spinosa*) and, in some parts of Britain, cherry-plum (*Prunus cerasifera*), while both species of hawthorn (*Crataegus monogyna* and *C. oxyacanthae*), flowering in early summer are perhaps the single most useful nectar plants of all. Some other flowering shrubs with similar open structure to their flowers, notably elder (*Sambucus nigra*), dogwood (*Cornus sanguinea*) and guelder rose (*Viburnum opulus*), do not seem to attract the diversity of invertebrates as does *Crataegus*. Later in summer bramble (*Rubus* spp.) and wild rose (*Rosa* spp.) are very valuable woody species (old dead *Rubus* stems are also valuable as nesting sites for some woodland solitary bees and wasps), while umbelliferous herbs such as hogweed (*Heracleum sphondylium*), angelica (*Angelica sylvestris*), composites such as golden-rod (*Solidago virguarea*), yarrow (*Achillea millefolium*) and even various thistles (*Cirsium* spp.) and ragworts (*Senecio* spp.) are very useful (Plate 10.1). Ivy (*Hedera helix*) is a valuable nectar source at the end of the year for species that overwinter as adults. One or two species of plant attract specialist nectar feeders, perhaps the most useful being various cinquefoils (*Potentilla* spp.), speedwells (*Veronica* spp.), bugle (*Ajuga reptans*), ground ivy (*Glechoma hederacea*) and figworts (*Scrophularia* spp.).

Deadwood

Initially, new woodland can offer very little to the fauna of saproxylic species — those associated in some way with dead and decaying timber. The more specialised species of this fauna may not benefit from current plantation woodland for hundreds of years. Many parks established in the 17th and 18th centuries, now with significant numbers of very old trees, appear still not to have developed the rich invertebrate faunas found in parks of medieval origin with similar appearance (K.N.A. Alexander, personal communication).

However, there are many common saproxylic species that do not require large or particularly old timber, and these include many attractive species such as hoverflies and longhorn beetles (Plate 10.3). The retention of thinnings and brashings from the first planting cycle, leaving standing failed plantings and the inclusion of non-intervention areas in the woodland where all deadwood, both standing and fallen, is allowed to remain, will all allow for colonisation by these species. Even fallen twigs support some of such species, especially if they are in large accumulations in moist shady conditions. 'Habitat piles' of brushwood and small branches may contribute to the development of

Plate 10.2. Many plant-feeding insects prefer young foliage of young trees, and will rapidly colonise new woodland, especially in sheltered areas on the edges of open spaces. The green sawfly (*Rhogogaster viridis*) feeds on the foliage of a wide variety of trees and shrubs (R.S. Key, English Nature)

Plate 10.3. Few deadwood species will colonise new woodland in the first half century or so. One that might is the wasp beetle *(Clytus arietis)*, which develops in much smaller pieces of deadwood, such as small branches, than the less common species. It needs open space with an abundant supply of flowering plants as an adult (R.S. Key, English Nature)

this fauna as the woodland develops, by maintaining important humidity, at least at the bottom of the pile. Larger branches and, eventually, fallen trees are best left where they fall on the woodland floor (Key, 1994a).

Water bodies in new woodland

The deliberate creation of pools in woodland may enhance the development of a rich invertebrate fauna, although the species concerned usually may have little particular association with woodland. Typical wetland species, however, are likely to benefit from the shelter and warmth provided by surrounding trees, provided that these are not so close as to cast shade or to cause problems from excessive leaf-fall into the water. Problems may arise through summer draw-down, although if this results in the creation of a seasonally flooded water body this may be to the advantage of scarcer, more specialised invertebrates that are unable to compete with the fauna typical of permanent water bodies. Very small temporary water features in woodland, particularly on clay, even down to the size of puddles, wheelruts and hoof prints, may be of value to this fauna (Bratton, 1990) but are easily lost in surfacing or grading operations. They ought not to be automatically regarded as negative features in woodland rides and tracks and eliminated.

While there is a small, but specialised invertebrate fauna associated specifically with shady pools or shady streams in woodland, these tend to be restricted to ancient woodland sites. It is probably better to concentrate on the creation of open-space water bodies for invertebrates in new woodland.

INTRODUCTION OF INVERTEBRATES INTO NEW WOODLAND

Many truly woodland species may either be very slow to colonise new woodland or not even colonise at all and it may, therefore, seem sensible to consider their deliberate introduction. Which species is it then appropriate to consider introducing? There may be thousands of species that inhabit older semi-natural woodland in the same area as new woodland and it is obviously impractical to consider introducing them all one-by-one. The ecology of many is so poorly known that in any case the likelihood of success is very low.

It is easy to be tempted to consider introduction only of those species that people find attractive. Usually this means the butterflies, but sometimes a few species of grasshopper and/or cricket and day-flying moths are also considered. This may be a laudable aim in woodland intended for human amenity and this can add greatly to the public enjoyment of the woodland. However, it carries with it the danger of encouraging complacent attitudes towards the fauna of truly ancient woodland. If a few of the more conspicuous elements are translocated successfully to new sites then the all-important public pressure to conserve these ancient sites may be lessened, despite the total fauna of less conspicuous species being perhaps two orders of magnitude richer.

The new woodland may also be considered as a suitable place for the transferral of species, or indeed whole elements of habitat such as soil or turves, from other areas of woodland or other habitat threatened with destruction. This again may seem to be a laudable aim, but carries with it similar dangers. There is an increasing tendency to see habitat creation or translocation, using existing ancient semi-natural habitat, as acceptable mitigation of the effects of development and destruction of ancient sites (Key, 1994b). This loses the historical continuity of that habitat, and often conspicuous, popular species, rather than those with exacting habitat requirements, are used as measures of success. Success in transferring whole invertebrate faunas, as opposed to individual species, has also been low, and is extremely unlikely for woodland. It has, however, rarely been adequately monitored. Such transplantation may, however, indeed help the colonisation of the new woodland with true woodland species, but such habitat should never be transferred more than a few miles, lest the ability to study the natural range of species in Britain be jeopardised.

It has been suggested that the importation of dead timber from sites with rich faunas into other woodland will facilitate the colonisation by deadwood species (Speight, 1989). This would certainly be inappropriate within the first several decades of the new woodland's existence, until the wood was producing sufficient dead woody material for colonisation.

Guidelines for the establishment of invertebrates are given by the Joint Committee for the Conservation of British Invertebrates (JCCBI, 1986) and the British Butterfly Conservation Society (BBCS, 1990), and should be followed in any establishment attempts.

SUMMARY AND CONCLUSION

This chapter is deliberately cautious in describing possible benefits to invertebrate conservation from the creation of new woodland. Low powers of dispersal of many woodland species means that the neediest species are the ones least likely to benefit, while deliberate introductions to augment invertebrate faunas are rarely a practical option. The defence and proper conservation management of existing old woodland should always take highest priority and the creation of new woodland is never mitigation against the loss of such old woodland.

However, new woodland does have a great deal to offer invertebrate conservation. Sympathetically positioned, it can considerably augment the fauna of patches of existing habitat of many types, especially existing woodland. Many species of invertebrate may be quick to colonise new plantations and some of these are attractive, easily observed species that may be enjoyed by human visitors to the wood. Encouragement of human enjoyment of these species through well-planned woodland layout, including the incorporation of suitable nectar plants, will benefit invertebrate conservation by increasing public sympathy towards organisms that are generally unappreciated. In the much longer term, as the new woodland matures, woodland invertebrates more in need of conservation measures begin to benefit. It is, therefore, vital that we pay adequate attention to the old woodland refuges that these species currently inhabit.

REFERENCES

Bratton, J.H. (1990). Seasonal pools — an overlooked invertebrate habitat. *British Wildlife*, **2**(1), 22–29.

BBCS (1990). *Butterfly Releases — British Butterfly Conservation Society Code of Practice.* BBCS, Quorn.

Carter, C.I. and Anderson, M.A. (1987). Enhancement of Lowland Forest Ridesides and Roadsides to Benefit Wild Plants and Butterflies. Forestry Commission Research Information Note 126.

Dawson, D. (1994). *Are Habitat Corridors Conduits for Animals and Plants in a Fragmented Landscape?* English Nature Research Reports, 94. English Nature, Peterborough.

Elton, C.S. (1966). *The Pattern of Animal Communities.* Methuen, London.

Foster, A.P. (1986) (unpublished). *Review of Invertebrate Sites in England: Norfolk Breckland.* Invertebrate Site Register Report No. 94. Nature Conservancy Council, Peterborough, 2 vols.

Fuller, R.J. and Warren, M.S. (1990). *Coppiced Woodlands: Their Management for Wildlife.* Nature Conservancy Council, Peterborough.

Hammond, P.M. (1974). Changes in the British Coleopterous fauna. In: Hawksworth, D.L. (ed.). *The Changing Flora and Fauna of Britain.* Proceedings of a Symposium of the Systematics Association at the University of Leicester, April 1973. Academic Press, London, 323–369.

Harding, P.T. and Rose, F. (1986). *Pasture-woodlands in Lowland Britain.* A review of their importance for wildlife conservation. Institute of Terrestrial Ecology, Huntingdon.

JCCBI (1986). *Insect Re-establishment — a Code of Conservation Practice.* Joint Committee for the Conservation of British Insects, London.

Kennedy, C.E.J. and Southwood, T.R.E. (1984). The number of species of insects associated with British trees: a re-analysis. *Journal of Animal Ecology*, **53**, 455–478.

Key, R.S. (1994a). Invertebrate conservation and dead wood. Section 6.2. In: *English Nature (1994). Species Conservation Handbook.* English Nature, Peterborough.

Key, R.S. (1994b). Invertebrates and mitigation — a word of caution. Section 7.5. In: *English Nature (1994). Species Conservation Handbook.* English Nature, Peterborough.

Key, R.S. and Gent, A. (1993). Bare but not barren. *Enact — Managing Land for Wildlife*, **1**, 15–16.

Kirby, P. (1987) (unpublished). *Review of Invertebrate Sites in England: Cambridgeshire.* Invertebrate Site Register Report No. 85. Nature Conservancy Council, Peterborough, 2 vols.

Kirby, P (1992). *Habitat Management for Invertebrates: A Practical Handbook.* Royal Society for the Protection of Birds, Sandy, Bedfordshire.

Lonsdale, D. (1991). Derelict land and wasteland. In: Fry, R. and Lonsdale, D. (eds). *Habitat Conservation for Invertebrates — A Neglected Green Issue.* The Amateur Entomologists' Society, Middlesex, 189–192.

Speight, M.C.D. (1989). *Saproxylic Invertebrates and Their Conservation.* Nature and Environment Series No. 42. Council of Europe, Strasbourg.

Spellerberg, I.F. and Gaywood, M.J. (1993). *Linear Features: Linear Habitats and Wildlife Corridors.* English Nature Research Reports, 60. English Nature, Peterborough.

Warren, M.S. (1987). The ecology and conservation of the heath fritillary butterfly, *Mellicta athalia.* II. Adult population structure and mobility. *Journal of Animal Ecology*, **24**, 483–498.

Warren, M.S. and Fuller, R.J. (1993). *Woodland Rides and Glades: Their Management for Wildlife.* Edition 2. Joint Nature Conservation Committee, Peterborough.

Warren, M.S. and Key, R.S. (1991). Woodlands: past, present and potential for insects. In: Thomas, J. and Collins, M. (eds). *Conservation of Insects and Their Habitats.* Proceedings of 15th Symposium of the Royal Entomological Society of London, 1989. Academic Press, London, 155–211.

Welch, R.C. and Greatorex-Davies, J.N. (1993). Colonization of two *Nothofagus* species by Lepidoptera in southern Britain. *Forestry*, **66**, 181–203.

11 Bird Populations in New Lowland Woods: Landscape, Design and Management Perspectives

R. J. FULLER, S. J. GOUGH and J. H. MARCHANT

INTRODUCTION

The supreme mobility of birds gives them the capacity to colonise new woodland more rapidly than most organisms. A striking example is the development of bird communities in polder woodland in the Netherlands. Over the last 50 years, substantial areas of the polders have been planted with conifers and mixed broadleaves. Many of these new woods are situated in areas otherwise devoid of trees; in some cases the nearest established woodland bird communities are more than 10 km away. Within just 20 to 30 years of planting, large numbers of bird species, and high densities of birds, have become established in many of these woods (Rob Bijlsma, personal communication; Bremer, 1980; Verstrael, 1989). The fast development of these Dutch woodland bird communities may be attributable also to the rapid structural development of much polder woodland. Rates of tree growth, for example, are extremely high.

The polders demonstrate that rich bird communities can develop quickly in woodland offering a suitable habitat. More widely, however, there is much variation in woodland bird communities, among both recently planted and established woods (Fuller, 1994). An understanding of the processes underlying this variation is important in attempting to create new woods that are, in themselves, rich habitats for birds and that enhance bird populations at a wider landscape level.

Attributes of individual woods are important in determining the potential suitability of any habitat patch for a species. For many woodland species, however, the dynamics of their populations will be influenced strongly by processes operating at a wider scale (e.g. Opdam, 1990; Dunning et al., 1992). This is certainly true of those many bird species where individuals that breed or roost within woods often feed in surrounding habitats, and where dispersal of individuals from one habitat to another appears to be frequent. Furthermore, some woods may act as population sources (Pulliam, 1988; Pulliam and Danielson, 1991) for bird species which are widely distributed across lowland landscapes. In general, far less is known about such potential landscape processes than about

The Ecology of Woodland Creation. Edited by Richard Ferris-Kaan.

how the characteristics of individual woods affect the bird communities that live within them (Fuller, 1995).

OBJECTIVES

Two major questions are addressed in this chapter.

- First, what is the contribution that woodland makes to bird communities in lowland English landscapes? This is especially relevant because various initiatives for habitat creation exist at the present time and some evaluation of the worth of woodland creation, as opposed to hedgerow creation for example, is desirable.
- Second, how can individual woods be designed and managed in such a way that they are likely to develop rich bird communities? In the latter context, we draw a distinction between woodland *design* and woodland *management*, both of which are relevant to birds. We focus mainly on the former, which embraces the features that are generally determined at the time of planting: woodland location, area and tree species composition. Management generally covers those aspects of stand treatment subsequent to planting.

STUDY AREAS, MATERIALS AND METHODS

This chapter draws on published studies and on unpublished censuses of breeding birds. The counts of birds were made with a territory mapping method in which locations and activities of birds were recorded on 1:2500 maps on several different visits in each breeding season (International Bird Census Committee, 1969; Marchant, 1983). The individual observations are termed 'registrations', and the spatial and temporal locations of these have been interpreted to assess numbers and distributions of territories within the study areas. Most of the analyses focus on songbirds (defined as all passerines excluding Corvidae and starling (*Sturnus vulgaris*)), because territory mapping is most suited to estimating numbers of these species. Details of the bird censuses used in this chapter are given below.

The Manydown Farm study

In 1984 British Trust for Ornithology (BTO) staff censused 562 ha of farmland and woodland at Manydown Farm, near Basingstoke, Hampshire. Data from this very large-scale census reveal broad patterns of bird distribution across different habitats at a landscape scale. The study area consisted of 473 ha of farmland, 64 ha of woodland and scrub and 25 ha of farmsteads, houses and gardens. The fields were predominantly arable. Only 11% of the field area was grass and the main crops were spring barley (175 ha), winter wheat (125 ha) and winter barley (98 ha). There were 17.2 km of hedgerow, 13.7 km of which were less than 1.5 m tall and frequently contained gaps, while the remaining 3.5 km were taller than 2.0 m and rarely contained gaps. Most hedges did not contain mature trees. The woodland existed in seven blocks and was mostly hazel (*Corylus avellana*) coppice (61% by area), at various stages of growth, under pedunculate oak (*Quercus robur*) standards. The remainder was high forest (28%), both mixed broadleaved/conifers and predominantly broadleaved, and scrub (11%). It seems likely that the efficiency of the census was lower in the woodland than in the other habitats due to the greater complexity of the vegetation, such that woodland densities were probably underestimated relative to those for open farmland.

Other BTO studies

Bird census data were also derived from other plots surveyed for the BTO's Common Birds Census (CBC) (Marchant *et al.*, 1990). The distribution of birds within eight farmland CBC plots in south and central England was examined to assess the relative occurrence of territories in hedges and in small woods at a plot scale. Individual territories were allocated to the various habitats of which the plot was composed. Territories straddling different habitats were divided between the habitats according to the proportion of their registrations in each. Each of the plots had been censused for 9 years or more during 1966–86 and each contained both hedges and small woods. Details of the plots are as follows:

1. Hertfordshire, mainly arable, 57 ha (1.3 ha woodland)
2. Dorset, mixed, 64 ha (3.8 ha)
3. Leicestershire, mixed farmland, 92 ha (2.5 ha)
4. Devon, mainly grass, 52 ha (9.6 ha)
5. Warwickshire, mixed farmland, 64 ha (1.2 ha)
6. Nottinghamshire, mainly arable, 74 ha (1.8 ha)
7. West Sussex, mainly grass, 58 ha (0.6 ha)
8. Buckinghamshire, mainly grass, 73 ha (1.5 ha).

To enable direct comparisons between woods and hedges, it was necessary to estimate the areas of land covered by hedges. No direct measurements of hedgerow width were available for the plots, so a width of 3 m was assumed. This was considered to be reasonable because measurement of a random sample of 27 hedge widths in Norfolk and Suffolk gave a mean width of woody vegetation of 2.7 m (range 1.5–5.0 m). Calculation of densities based on area for narrow linear features such as hedges is problematical for two main reasons. First, hedgerow birds may defend a given length of hedge unlike woodland birds, which are more likely to defend an area. Second, some hedgerow territories may extend onto adjacent farmland by an unknown amount, and the birds may also use it for feeding. This makes it difficult to assess the appropriate habitat area for use in calculating density of birds using hedgerows. Although the densities calculated in this way are potentially misleading if used uncritically (Haila, 1988), this approach does allow broad differences in patterns of habitat selection shown by different species to be identified. For individual bird species, in the above 6 years, selection or avoidance of woodland was calculated by Jacobs' (1974) Preference Index (*D*):

$$D = (r-p)/(r+p-2rp)$$

where *r* is the proportion of territories in woodland, and *p* is the proportion of total woody area (i.e. woodland plus hedgerow) contributed by woodland. *D* potentially ranged from +1 (total selection of woodland) to −1 (total selection of hedges). Mean values of D were calculated across years. (D was not calculated for any species-year combination where there were fewer than 10 territories.)

The CBC also provided a sample of woods which allowed a preliminary exploration of relationships between population stability and woodland size. The plots chosen were broadleaved isolates (parts of larger woodland areas were excluded) censused for at least 12 consecutive years and which had not undergone major habitat change during that period. Active coppice, and other woods with substantial areas of pre-thicket growth, were excluded.

BIRD DISTRIBUTION WITHIN LOWLAND LANDSCAPES

Woodland is one of several vegetation types that constitute lowland agricultural land-scapes in western Europe. The effects of landscape composition — the relative amounts of different habitats (Dunning *et al.*, 1992) — on bird populations have hardly been studied. Some insight into the contribution that woodland makes to the bird populations living within lowland landscapes can, however, be gained from the Manydown census.

The woodland at Manydown contributed 53% of the songbird territories but it covered only 11% of the study area. The fields and hedges covered 84% of the area but they contributed only 26% of the songbirds. The overall densities of songbirds (as territories km^{-2}) in different habitats at Manydown were: woodland 969, habitations 867, fields 21. The overall hedgerow density was 13 territories km^{-1} or, assuming a hedge width of 3 m, 4279 territories km^{-2}. The hedges at Manydown appeared to support more than four times as many birds per unit area than the woodland but this finding should be interpreted cautiously. Hedgerow nesting birds may be making far more use of the surrounding land-scape than birds in woodland and some hedgerow territories may include areas of adja-cent land. This means that our method of calculating density will have overestimated hedgerow densities relative to those of woodland; this effect will have been further exac-erbated by the higher efficiency of the hedgerow censuses (see above).

There were large differences in bird species composition between the woodland and hedges at Manydown (Figure 11.1). Several species were virtually confined to the wood-land, including nuthatch (*Sitta europaea*), treecreeper (*Certhia familiaris*), willow tit (*Parus montanus*), marsh tit (*Parus palustris*), goldcrest (*Regulus regulus*), willow warbler (*Phylloscopus trochilus*), blackcap (*Sylvia atricapilla*) and garden warbler (*Sylvia borin*). In contrast, only whitethroat (*Sylvia communis*) was a strong hedgerow specialist, but yellowhammer (*Emberiza citrinella*) and dunnock (*Prunella modularis*) were also associated with hedges rather than with woods. The species marked as 'ubiqui-tous' in Figure 11.1 were common in both woods and hedges. Patterns of habitat selec-tion are further explored using CBC data in Figure 11.2, for six species that are widespread in both hedges and woodland. Blackbird (*Turdus merula*) and dunnock clearly preferred hedges. However, chaffinch (*Fringilla coelebs*), wren (*Troglodytes troglodytes*), blue tit (*Parus caeruleus*) and robin (*Erithacus rubecula*) showed no consis-tent preference. At the level of major groups of songbirds, there were striking differences in habitat selection. Finches and buntings were predominantly associated with hedges, perhaps because these mainly seed-eating species do much of their feeding in open coun-tryside. No consistent pattern emerged for tits, but warblers strongly selected woodland. The bird communities in hedgerows appear to be similar to those found in very small

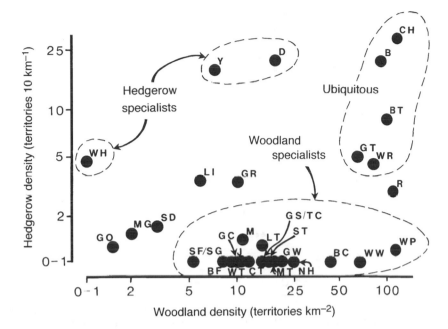

Figure 11.1. Comparison of hedgerow and woodland bird communities at Manydown Farm, Hampshire, England, 1984. Species primarily associated with hedgerows and woodland are indicated. Note that although mistle thrush (*Turdus viscivorus*), starling (*Sturnus vulgaris*) and woodpigeon (*Columba palumbus*) mainly nest within woodland, they are likely to make much use of farmland for feeding. Species symbols are defined in the appendix

patches of woodland and scrub (< 0.1 ha). In East Anglia species such as dunnock, blackbird, chaffinch and yellowhammer will breed in these small patches, but warblers, treecreeper, marsh tit and nightingale (*Luscinia megarhynchos*) avoid them (P. Bellamy, S. Hinsley, personal communication). Moreover, McCollin (1993) has suggested that finches and buntings are more characteristic of small than large woods.

The results from Manydown indicate that, at a landscape level, woodland can support a disproportionately high number of the breeding songbirds, and that a higher number of species are dependent on woodland than on hedgerows. It is possible, however, that where hedges carry many mature trees, the difference between hedgerow and woodland bird communities would be less marked than was the case at Manydown. Nonetheless, the patterns of habitat distribution indicated in Figure 11.1 probably broadly reflect the situation over large areas of lowland Britain. The CBC data indicate that woodland is generally more strongly selected by warblers and this is consistent with the Manydown findings. Densities of several species of warblers are certainly higher in woodland and scrub than in hedgerows, but there are two exceptions: whitethroat (*Sylvia communis*) and lesser whitethroat (*Sylvia curruca*), which are common breeders in hedges. The CBC data show that species vary greatly in habitat selection from one farm to another. This was presumably due to variation in habitat quality offered by the individual hedges and woods.

These results indicate that several species are more strongly associated with woodland than with other habitats, and thus that woodland contributes substantially to the avian

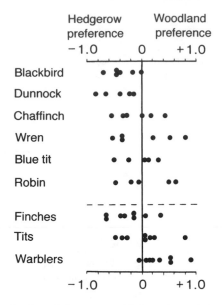

Figure 11.2. Habitat selection of six common bird species and three groups of songbirds. Each dot represents a different farmland Common Bird Census (CBC) plot. Preference for hedgerow (−1) or woodland (+1) is shown by the Jacob's Preference Index. Eight farms were studied, but sample sizes were too small to allow calculation of the index for all species on all farms. The bird groups are: finches – Fringillidae and Emberizidae; tits – Paridae; warblers – Sylviidae. See Methods section for more details

richness of lowland landscapes. These 'snapshot' censuses cannot, however, tell the full story. Densities do not necessarily indicate which are the optimal habitats, partly because breeding success is not necessarily correlated positively with population density (van Horne, 1983; Vickery *et al.*, 1992). Moreover, species occupying a wide range of habitats can show dynamic patterns of habitat selection whereby their distribution across habitats changes with overall population level (e.g. O'Connor, 1986; Rosenzweig, 1991). Censuses conducted in years when populations are high may generate misleading conclusions about habitat preferences. Where such processes operate, numbers are expected to be more stable in the preferred habitats because vacancies appearing there will quickly be filled by incoming birds from elsewhere.

Preliminary analyses of long-term distribution of breeding songbirds within 15 farmland CBC plots suggest that three species — wren, blackcap and willow warbler — frequently show a preference for woodland over hedgerows. When their population levels were low, territories of these birds were centred on woodland, but the proportion of territories in hedges increased when overall population levels were high. An example for blackcap is shown in Figure 11.3. Similarly, Osborne (1982) has shown that chiffchaff territories on a Dorset farm were associated mainly with small woods, and that hedges distant from such woods were occupied only in years when the overall population level was high.

As outlined earlier, some woods may act as population sources by producing a surplus of young which disperse and settle in other patches or habitats, perhaps including hedges. Adequate tests of this idea require measurement of productivity and survival in different

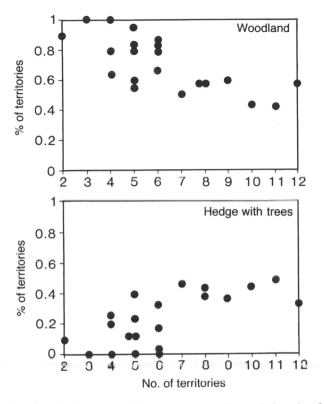

Figure 11.3. Relationship between habitat occupancy and population size for blackcaps (*Sylvia atricapilla*) on a farm in Hertfordshire, southern England, over a 21-year period (1967–87). Individual data points are years. When population levels are low, territories are concentrated in the woodland but, as numbers increase, the species colonises hedges with trees

habitats, coupled with an understanding of the spatial dynamics of distribution. Work at this level of detail does not appear to have been carried out for any bird species in lowland European landscapes. It has been suggested that hedgerows are suboptimal habitats for many birds (Murton and Westwood, 1974), the implication being that their bird populations either 'overflow' from other habitats, or are maintained by immigration from source populations in other habitats, especially woodland. This is a simplistic generalisation. Individual woods and hedges vary enormously in their structure and floristics and, therefore, in the quality of habitat they offer birds (O'Connor and Shrubb, 1986; Fuller, 1995). It is entirely feasible that both woodland and hedgerow can at times act as either sources or sinks, depending on the species concerned and on the quality of the individual patches. It seems appropriate to regard these as complementary habitats for, as shown above, they differ substantially in the make-up of their bird communities. Furthermore, woodland may be potentially richer in bird species, but hedges support huge numbers of birds and they make an enormous contribution to the bird life of lowland landscapes, especially where woodland is scarce.

IMPLICATIONS OF WOODLAND DESIGN

The structure and composition of bird communities in woodland are influenced by a large number of interacting factors (Fuller, 1995). Those factors with special relevance for new woodland are outlined below.

Woodland location and landscape composition

The bird community that develops within a new wood might be affected by the composition of the surrounding landscape. With respect to new woodland, pertinent questions are whether new woods should be created in close proximity to existing ones, and whether they should be linked by connecting features. Knowledge of the spatial dynamics of birds in fragmented landscapes lags far behind the development of metapopulation theory (see Spellerberg, Chapter 4). One notable exception stems from research on nuthatches in the Netherlands (Verboom et al., 1991). This species lives in mature broadleaved and mixed woodland but in many farmland landscapes not all the suitable habitat is occupied at any one time, and its distribution is dynamic. Extinction rate is highest in small patches, which carry relatively few breeding pairs, and in patches of poor habitat quality. On the other hand, nuthatches most readily colonise unoccupied patches of woodland that are in the vicinity of occupied habitat.

The colonisation of new woodland depends on dispersal. Indeed, dispersal is important in ensuring the persistence of populations in many woodland patches (Opdam, 1990). Where a patch forms a population sink, the population cannot persist in the absence of immigration. However, local extinctions will also occur in more favourable habitat, for purely stochastic reasons. The probability of colonisation or recolonisation will be linked both to the innate dispersal ability of the species, and to the size of the population in the surrounding region which, in part, will be determined by the amount of suitable habitat. There has been no formal comparison of patterns of natal or breeding dispersal of different woodland bird species in Britain. Nonetheless, it is clear that there is considerable variation between species in their local movements. A small number of species, including marsh tit and nuthatch, are highly sedentary. For these species it is quite possible that the isolation of a wood from other potential habitat may determine whether colonisation occurs and, if it does, whether a population persists. At the other extreme, there appears to be substantial year-to-year turnover of territory-holding individuals of several common species in small woods, implying much movement of individuals within the landscape, perhaps coupled with the existence of large 'floating populations' (Hinsley et al., 1992; and personal communication). Among these species are wren, robin, chaffinch, dunnock, blackbird, great tit and blue tit.

From work in the Netherlands it appears that both numbers and types of bird species occurring in farm woodlands are related to their isolation from other woodland (Opdam et al., 1985; van Dorp and Opdam, 1987). In particular, the occurrence of species associated with mature woodland decreased in more isolated sites. Such species included woodpeckers, treecreeper, nuthatch and marsh tit. It is possible that these so-called 'forest-interior' species are less likely to colonise woods that are isolated from others because they tend not to disperse over large distances. Under such circumstances it is possible that some individual woods may act as stepping stones rather than as sources for colonisation. Further evidence of isolation effects comes from studies of the distribution of birds among woods in eastern England (S. Hinsley and P. Bellamy, unpublished). The presence of several species of breeding birds in these woods was related to the amount of hedgerow and woodland in the surrounding landscape, or to the number of hedges connected to a wood. The ecological function of features that connect different habitat patches, such as hedgerows between farm woods, is controversial (Hobbs, 1992; Taylor *et al.*, 1993). In particular, it is unclear whether these linear features act as corridors aiding dispersal. Alternatively, connecting features might simply act as additional habitat, thereby contributing to the pool of potential colonists. There is no strong evidence that hedgerows do act as corridors for woodland birds.

In summary, the few results available to date suggest that proximity to existing woodland may be a significant factor in the colonisation of woods by some bird species, especially sedentary ones. In general, however, location of new woodland with respect to existing woodland and hedgerows is likely to be a more important consideration for less mobile animals, such as several groups of invertebrates (Fuller and Warren, 1991; Key, Chapter 10).

Woodland size

A disproportionate amount of work has focused on relationships between woodland size and natural communities (see Spellerberg, Chapter 4). This section briefly summarises key points that have emerged as being of special relevance to birds in new woodland. Four attributes of bird communities appear to be related to woodland area:

- Woodland area is a strong predictor of the number of bird species found in a wood (Moore and Hooper, 1975; Woolhouse, 1983; Ford, 1987; van Dorp and Opdam, 1987; Hinsley *et al.*, 1992).
- Area of woodland is related to the composition of associated bird communities. Bird communities in small woods are not random samples of those found in large woods (Nilsson, 1986). In Britain, it seems that several species tend to avoid the smallest woods, for example nightingale, chiffchaff, marsh tit and jay (*Garrulus glandarius*) (S. Hinsley, personal communication). On the other hand, no bird species are confined to small woods.
- In contrast to the relationship with species number, woodland area is generally inversely related to the overall density of breeding birds (Nilsson, 1986; Ford, 1987). This is a consequence of edge effects because densities of birds are commonly higher at the edge than in the interior of woodland (Fuller, 1991).
- The stability of bird populations in very small woods (<2 ha) appears to be lower than in larger woods and the annual turnover of individual adult birds in very small woods can be high (Hinsley *et al.*, 1992).

Some landowners may be in the position to choose between planting several small or fewer large woods of equivalent area. Which strategy is best? Two British studies have addressed this issue from the standpoint of maximising the numbers of breeding bird species in lowland woods (Ford, 1987; Woolhouse, 1987). Both concluded that one large wood will support fewer species than several small woods of equivalent total area. However, Woolhouse (1987) suggested that in the absence of immigration (admittedly an unlikely situation for most woodland birds in Britain, with the possible exception of some sedentary species), a single large wood would eventually retain more species than several isolated small woods. Woolhouse (1987) recommended the best strategy was to go for several small woods. This may be appropriate on the scale on which he was working (he compared a single 40 ha wood with five 8 ha woods), but there are potential disadvantages in advocating this approach for much smaller woods. There would seem to be several reasons for avoiding a planting strategy which created *only* very small woods (say < 2 ha), especially isolated ones. First, several species avoid the smallest woods (see above). Second, rates of predation may be relatively high in small woods because predators such as crows (*Corvus* spp.), which are mainly associated with the surrounding farmland, may inflict high nest losses at the woodland edge (Andrén and Angelstam, 1988). Third, the populations of most species in such small woods often consist of just one or two territories. For stochastic reasons, one would expect these small populations to be more prone to extinctions than populations in larger woods.

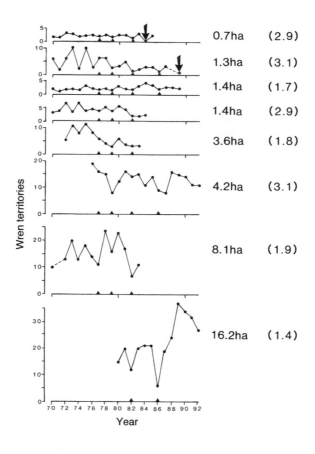

(a)

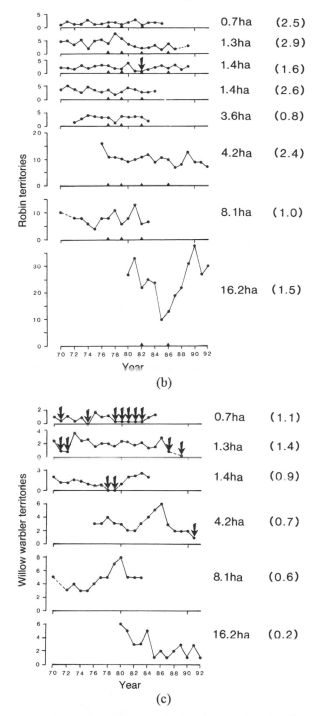

(b)

(c)

Figure 11.4. Numbers of breeding birds (territories) in woods of different sizes: (a) wren (*Troglodytes troglodytes*); (b) robin (*Erithacus rubecula*); (c) willow warbler (*Phylloscopus trochilus*). In the case of wren and robin, breeding seasons preceded by a severe winter are marked with a triangle. Years when less than one complete territory was recorded within the wood are marked with an arrow. Mean density (territories ha⁻¹) is given in parentheses next to the area of each wood

Hinsley *et al.* (1992) showed that year-to-year turnover of species was much greater relative to the total number of species present in small woods but we know of no studies that have compared long-term population stability and, more to the point, population persistence, in woods of different sizes. The long-term fluctuations of three species shown in Figure 11.4(a),(b),(c) for woods varying in size are, therefore, of interest, even though the sample sizes are very small. The two resident species — wren and robin — were chosen because they fluctuate strongly in relation to severity of winter weather. The third species, willow warbler, is a long distance migrant which is unaffected by overwinter conditions in Britain. As expected, population size of each species was generally related to woodland size, but mean population density tended to be higher in the smallest woods. The amplitude of population fluctuation was not obviously related to woodland size; populations appeared no more stable in large than small woods. Indeed, the most pronounced responses of wren and robin to the severe winters in the 1980s were in the largest woods, though it is notable that these populations never approached extinction. There was weak evidence that populations were less persistent in the smallest woods. Willow warbler became extinct in each of the three smallest woods but not in the three largest. Wren became extinct in the smallest wood, but not in any of the large woods. There were, however, no extinctions of robin. Counts, of course, give an incomplete assessment of stability because they tell us nothing about the turnover of individual birds. The apparent stability of numbers of wren and robin in the smallest woods in Figure 11.4(a) and (b) may conceal a much higher turnover in territory holders than in the larger woods. Populations in small woods may only appear persistent because of constant immigration.

Small woods (< 2 ha) are, however, by no means devoid of value for birds. They can support a wide range of species and may actually be favoured by some species, including seed-eating finches and buntings (McCollin, 1993). Small woods are also often especially suitable for those species such as turtle dove (*Streptopelia turtur*) and warblers which are associated with the edges of larger, mature woods (see below). Many landowners planting new woods will be aiming to establish high quality gamebird habitat. In such cases there is a strong argument for creating a preponderance of small woods because free-living pheasants (*Phasianus colchicus*) are woodland edge birds. The amount of woodland edge is a major predictor of pheasant densities in lowland landscapes (Robertson, 1992; Robertson *et al.*, 1993a,b). Large woods can, however, be improved for pheasants through suitable management, involving the creating of open space through coppicing, skylighting or wide rides (Bealey and Robertson, 1992). For the benefit of bird populations in general, however, Ford's (1987) recommendation that landscapes should hold woods of a variety of sizes should be applied.

Tree species composition

The tree species of a new wood will be determined largely, but not irreversibly, at the time of planting. Rich woodland bird communities are often thought to be a consequence of complex vegetation structure rather than of the tree species composition. This is a simplification of how birds respond to their woodland habitats. Structural complexity is certainly important but there is increasing evidence that floristics, at least at the level of trees and shrubs, is relevant to bird communities in the following ways (Fuller, in press; Fuller, 1995). Different species of trees offer birds different types of foods and different types of nest sites. Insectivorous birds, for example, prefer to feed in some species of trees more than in others, with each bird species differing in its preferences (Holmes and Robinson, 1981; Peck, 1989). There are differences also amongst frugivorous bird species in their preferred fruits (Snow and Snow, 1988). Furthermore, because the fruiting periods of different trees and shrubs may be staggered, plant diversity will affect food availability to some extent. Each tree species differs in its structural characteristics — both in foliage profile and in microhabitats such as bark texture and availability of holes — so woods that are floristically diverse tend to have complex structures. It is beyond the scope of this chapter to discuss individual tree species and mixtures of trees and their implications for birds; in fact, adequate information to do this thoroughly is unavailable. We are more concerned to draw attention to some general design features which should enhance the use of new woods by birds throughout the year.

Rather few species are confined either to broadleaved or to coniferous woodland, and both types of woodland can support rich bird communities. In the lowlands, however, there are rather more broadleaved than coniferous specialist bird species. Among the former are lesser spotted woodpecker (*Dendrocopus minor*), nuthatch, marsh tit, willow tit, nightingale and hawfinch (*Coccothraustes coccothraustes*). The main conifer specialists are coal tit (*Parus ater*) and goldcrest (*Regulus regulus*), but these will nest in broadleaves, albeit at lower density. Woods that are composed of a mixture of locally native broadleaved species are likely to be attractive to a wide range of birds. However, mixtures of broadleaves and conifers potentially carry very rich bird communities, though this has been little studied. The addition of even small amounts of broadleaved trees to otherwise pure coniferous stands can increase the densities of several species of birds (Bibby *et al.*, 1989). As a general rule, new woods should contain a variety of trees, including ones that are native to the area. It is also desirable that these should include both shrub species and canopy species because this will help to promote structural complexity.

MANAGEMENT OPPORTUNITIES IN NEW WOODS

Woodland management can profoundly alter the structure of woodland, both in terms of foliage complexity and canopy openness. The management system employed in any particular wood can be a dominant influence on the nature of the bird community (Fuller, 1995). Nonetheless, the implications for birds of different options and approaches to management once the stand has matured are not discussed here. Rather, the aim is to discuss those opportunities for enhancing bird communities through management strategies that are specially relevant to new woods.

Development of bird communities

There is much variation from one wood to another in the development of bird communities. Some new woods planted on farmland can be almost devoid of birds for up to 10 years (unpublished CBC data). Where the developing woodland has a vigorous field and shrub layer, particularly of bramble (*Rubus fruticosus*) and regenerating shrubs, bird colonisation is likely to be far more rapid. There are two approaches to speeding up the process. First, shrubs and trees which grow rapidly can be planted (see Harmer and Kerr, Chapter 8).

Depending on soil type these might include hawthorn (*Crataegus monogyna*), birch (*Betula* spp) and blackthorn (*Prunus spinosa*); some landowners will be deterred by its invasive nature, though blackthorn scrub can be a rich songbird habitat. The edges of the wood are particularly suitable areas to plant these shrubs (see below). Second, planting can be supplemented with natural regeneration (see Rodwell and Patterson, Chapter 5; Harmer and Kerr, Chapter 8); in many cases a very adequate woody cover can become established through natural regeneration alone. It is possible to take advantage of the fact that the seeds of many shrubs and trees are dispersed by birds. Natural regeneration can be enhanced by providing perches for birds, in the form of branches or poles (McClanahan and Wolfe, 1993). Where management for pheasants is a high priority, the rapid establishment of vegetation cover is important and Robertson (1992) describes techniques for achieving this.

The woodland edge

In many established, mature woods, several species of animals are associated with the external edges (Fuller, 1991; Fuller and Warren, 1991). The overall density and richness of bird communities is often highest at the external edge (Plate 11.1). Birds strongly associated with the outer woodland edge typically include: garden warbler, blackcap, willow warbler, chiffchaff, dunnock, wren, chaffinch, blackbird and long-tailed tit. Unlike North American forests (e.g. Askins *et al.*, 1987), lowland British woods hold no bird species that avoid the external edges of woods and which can be regarded as true forest-interior species. These strong edge effects appear to be underpinned by five interrelated factors.

- The vegetation structure at the edge is often more complex, in particular the shrub layer is frequently denser.
- The variety of shrubs and trees is often greatest at the woodland edge.
- Many shrubs flower best at the woodland edge and produce larger amounts of fruit there for birds to eat.
- Many insects appear to occur in large numbers at the woodland edge.
- The proximity to farmland is important to some species, for example pheasant, which breeds at higher densities along edges bordering arable rather than grass, the former being a better feeding habitat (Robertson, 1992).

Internal edges, such as those along rides, are not generally so attractive to birds, presumably because long-established external edges have gradually developed complex structures and communities of plants in response to constancy of light conditions. For many birds the advantages presumably outweigh any increased risks of predation at the woodland edge (Andrén and Angelstam, 1988).

Planting native shrubs at the edge, and encouragement of natural regeneration, will aid the rapid establishment of rich edge habitats. One strategy might be to leave an unplanted belt up to 20 m wide around the planted area within which scrub and woodland would be given the opportunity to develop through natural regeneration. In this way, an ecotone may be created grading from open land through scrub to mature woodland but, without fencing, this would only work where there were few deer (see Gill *et al.*, Chapter 13). In the longer term, the edges are appropriate areas for special treatments aimed at enhancing the wildlife of small woods. There is much to be said for maintaining the edge as a belt of scrub, perhaps 5 m to 10 m wide. This could be cut in fragments on a long rotation. The development of dense scrub would benefit warblers, and many insects too (see Key,

Plate 11.1. The edges of old established woodlands often hold larger numbers of breeding birds and a wider range of species than the interior. Factors often contributing to the richness of bird life at such edges include a high diversity of tree and shrub species and a complex vegetation structure. It is possible to accelerate the development of such features in new woodland by planting mixtures of native tree and shrub species at the edge (R.J. Fuller, British Trust for Ornithology)

Chapter 10), as well as providing a roost habitat for thrushes, finches and buntings which feed on surrounding farmland (Plate 11.1). There are many benefits to birds and other wildlife in allowing a scrub-dominated edge to develop, but another approach is to conduct a particularly heavy thinning at the edge of the wood (Pietzarka and Roloff, 1993). Opening up the canopy may promote a denser understorey, though this type of treatment would not be advisable on exposed sites where there was risk of windthrow.

CONCLUSIONS AND RECOMMENDATIONS

This review has aimed to identify approaches to woodland creation which will enhance bird communities both at the scale of the landscape and that of the individual wood. However, substantial gaps in knowledge have been identified which hamper the provision of firm advice. It is important, therefore, that these conclusions should be regarded as preliminary.

Some species of birds are associated far more strongly with woodland than with other habitats. The implication of the results presented above is that, in landscapes where woods are scarce, the addition of rather small amounts of woodland may contribute disproportionately to the overall numbers of birds and to avian richness. Furthermore, it seems probable that woods play a significant role in the dynamics of many bird populations at a landscape level.

If an aim of new woodland is to create a rich wildlife habitat, including birds, then it is important to consider the landscape context (see Bell, Chapter 3), as well as the attributes of the new wood itself. Where existing woodland and hedgerows are sparse, new woodland might appear to have the greatest potential to enhance landscape avian diversity. However, the creation of widely scattered, very small patches of woodland (< 2 ha) in

Plate 11.2. Natural birch regeneration within a broadleaved plantation. The development of unplanted shrubs and bramble within recently planted areas is to be welcomed for it can lead to a vigorous low shrubby growth which provides the cover required by several bird species, especially warblers (R.J. Fuller, British Trust for Ornithology)

these situations may merely lead to unstable bird populations within the woods, whose existence depended on constant immigration. Furthermore, sedentary woodland bird species may never reach such woods and, if they do, are unlikely to establish persistent populations. Especially where existing woodland is scarce, perhaps the best strategy for the long-term development of rich bird communities would be through the creation of moderately large blocks of woodland, larger than 2 ha, but preferably at least 5 ha. It would also be desirable to create clusters of such woods, though this may be unrealistic. The larger the woodland the less the risk of population extinction (see Spellerberg, Chapter 4). We suggest that the real value of creating relatively large patches of woodland lies in population persistence, rather than in species–area relationships *per se*.

This does not imply that planting small blocks of woodland is misguided. Their potential value as pheasant habitat has been acknowledged, and for this reason small woods are bound to be favoured by many landowners. Furthermore, the relatively large amounts of edge habitat created in small woods may be beneficial to warblers and other birds associated with shrubby and edge habitats. In landscapes with moderate or substantial amounts of extant woodland and hedgerow, modest additions of woodland may increase the overall amount of habitat available to birds such as woodpeckers, marsh tits and nuthatches, thus leading to larger populations of woodland birds in these districts.

In practice, the great majority of new woods will be plantations of timber trees or of short-rotation coppice. Though of no commercial value, except as cover for gamebirds, scrub has a place in the new woodland. It is patchily distributed in the lowlands yet it carries distinctive communities of breeding birds, often including high densities of summer visitors, forms an important roost habitat for birds, and can be rich in invertebrates. There is much scope for incorporating scrub into new woods, either by planting or by natural regeneration, especially at the woodland edge (Plate 11.2). Where small-scale planting is proposed in relatively treeless, open landscapes, the establishment of scrub may be an especially appropriate option (Plate 11.3).

Plate 11.3. Hawthorn scrub colonising a hillside in Buckinghamshire. This scrub is still too open to support a rich breeding bird community but in a few years, when the canopy is starting to close, it will be attractive to a variety of warblers and resident bird species. The bushes will also provide a source of winter food for berry-feeding thrushes. The establishment by natural regeneration of patches or belts of scrub may be feasible within, or at the edge of, those new woods where deer are not too much of a problem. The process can be aided by the provision of perches for birds (R.J. Fuller, British Trust for Ornithology)

Based on current knowledge, broad recommendations concerning the design and management of new woodland are as follows. These recommendations are made with native woodland bird species in mind.

- Plant a mixture of large and small woods, preferably with some exceeding 5 ha.
- Avoid creating very isolated woods, especially small ones (< 2 ha). Where possible, create clusters of woods, especially where small woods are being planted.
- Plant a variety of tree and shrub species native to the district, even where the crop is predominantly coniferous or composed of non-native trees.
- Plant underwood or shrub species as well as canopy species.
- Use natural regeneration to supplement planting. Encourage natural regeneration by providing perches for birds.
- Where possible, create and maintain areas of scrub. The edges of woodland are especially suitable for this.

Acknowledgements

We thank the following who censused the CBC plots: Keith Atkin, R. C. Branwhite, A. J. Clarke, D. A. Cohen, B. Constable, R. G. Gibbs, Miss Beatrice Gillam, Mrs G. D. Harthan, Brigadier J. A. Hopwood, Mrs A. Hughes, Leicestershire and Rutland Ornithological Society, W. E. Merrill, S. C. Nichols, Miss M. E. Price, Miss E. M. P. Scott, J. R. Spencer, Dr R. Stanford, W. Stuart-Best, Mrs J. Walton and Pip and Eve Willson. The work at Manydown Farm was funded by The Game Conservancy Trust as part of their Cereals and Gamebirds Project and was carried out by Dr Kenneth Taylor, Phil Whittington, JHM and RJF. We thank Mr H. R. Oliver-Bellasis, Dr Raymond O'Connor and Dr M. R. W. Rands for their help and advice during the work at Manydown. We are grateful to Dr Shelley Hinsley, Paul Bellamy and Dr Pete Robertson for their comments. Shelley Hinsley and Paul Bellamy of the Institute of Terrestrial Ecology kindly made some of their unpublished data available. This work was partly funded by the Joint Nature Conservation Committee (on behalf of English Nature, the Countryside Council for Wales and Scottish Natural Heritage).

REFERENCES

Andrén, H. and Angelstam, P. (1988). Elevated predation rates as an edge effect in habitat islands: experimental evidence. *Ecology*, **69**, 544–547.

Askins, R.A., Philbrick, M.J. and Sugeno, D.S. (1987). Relationship between the regional abundance of forest and the composition of forest bird communities. *Biological Conservation*, **39**, 129–152.

Bealey, C.E. and Robertson, P.A. (1992). Coppice management for pheasants. In: Buckley, G.P. (ed.). *Ecology and Management of Coppice Woodlands*. Chapman and Hall, London, 193–210.

Bibby, C.J., Aston, N. and Bellamy, P.E. (1989). Effects of broadleaved trees on birds of upland conifer plantations in north Wales. *Biological Conservation*, **49**, 17–29.

Bremer, P. (1980). Broedvogels van het Kuinderbos. *Vogeljaar*, **28**, 287–291.

Dunning, J.B., Danielson, B.J. and Pulliam, H.R. (1992). Ecological processes that affect populations in complex landscapes. *Oikos*, **65**, 169–175.

Ford, H.A. (1987). Bird communities on habitat islands in England. *Bird Study*, **34**, 205–218.

Fuller, R.J. (1991). Effects of woodland edges on songbirds. In: Ferris-Kaan, R. (ed.). *Edge Management in Woodlands*. Forestry Commission Occasional Paper 28. Forestry Commission, Edinburgh, 31–34.

Fuller, R.J. (1995). *Bird Life of Woodland and Forest*. Cambridge University Press, Cambridge.

Fuller, R.J. (in press). Relating birds to vegetation: influences of scale, floristics and habitat structure. Proceedings of the 12th International Conference of IBCC and EOAC, The Netherlands.

Fuller, R.J. and Warren, M.S. (1991). Conservation management in ancient and modern woodlands: responses of fauna to edges and rotations. In: Spellerberg, I.F., Goldsmith, F.B. and Morris, M.G. (eds). *The Scientific Management of Temperate Communities for Conservation*. British Ecological Society Symposium No. 31. Blackwell Scientific Publications, Oxford, 445–471.

Haila, Y. (1988). Calculating and miscalculating density: the role of habitat geometry. *Ornis Scandinavica*, **19**, 88–92.

Hinsley, S., Bellamy, P. and Newton, I. (1992). Habitat fragmentation, landscape ecology and birds. Institute of Terrestrial Ecology Report 1991–1992, 19–21.

Hobbs, R.J. (1992). The role of corridors in conservation: solution or bandwagon. *Trends in Evolution and Ecology*, **7**, 389–392.

Holmes, R.T. and Robinson, S.K. (1981). Tree species preferences of foraging insectivorous birds in a northern hardwoods forest. *Oecologia*, **48**, 31–35.

International Bird Census Committee (1969). Recommendations for an international standard for a mapping method in bird census work. *Bird Study*, **16**, 249–255.

Jacobs, J. (1974). Quantitative measurement of food selection. *Oecologia*, **14**, 413–417.

Marchant, J.H. (1983). *BTO Common Birds Census Instructions*. British Trust for Ornithology, Tring.

Marchant, J.H., Hudson, R., Carter, S.P. and Whittington, P. (1990). *Population Trends in British Breeding Birds*. British Trust for Ornithology, Tring.

McClanahan, T.R. and Wolfe, R.W. (1993). Accelerating forest succession in a fragmented landscape: the role of birds and perches. *Conservation Biology*, **7**, 279–288.

McCollin, D. (1993). Avian distribution patterns in a fragmented wooded landscape (North Humberside, UK): the role of between-patch and within-patch structure. *Global Ecology and Biogeography Letters*, **3**, 48–62.

Moore, N.W. and Hooper, M.D. (1975). On the number of bird species in British woods. *Biological Conservation*, **8**, 239–250.

Murton, R.K. and Westwood, N.J. (1974). Some effects of agricultural change on the English avifauna. *British Birds*, **67**, 41–69.

Nilsson, S.G. (1986). Are bird communities in small biotope patches random samples from communities in large patches? *Biological Conservation*, **38**, 179–204.

O'Connor, R.J. (1986). Dynamical aspects of avian habitat use. In: Verner, J., Morrison, M.L. and Ralph, C.J. (eds). *Wildlife 2000: Modeling Habitat Relationships of Terrestrial Vertebrates*. University of Wisconsin Press, Madison, 235–240.

O'Connor, R.J. and Shrubb, M. (1986). *Farming and Birds*. Cambridge University Press, Cambridge.

Opdam, P. (1990). Dispersal in fragmented populations: the key to survival. In: Bunce, R.G.H and Howard, D.C. (eds). *Species Dispersal in Agricultural Habitats*. Belhaven Press, London, 3–17.

Opdam, P., Rijsdijk, G. and Hustings, F. (1985). Bird communities in small woods in an agricultural landscape: effects of area and isolation. *Biological Conservation*, **34**, 333–352.

Osborne, P. (1982). The Effects of Dutch Elm Disease on Farmland Bird Populations. DPhil thesis, University of Oxford.

Peck, K.M. (1989). Tree species preferences shown by foraging birds in forest plantations in Northern England. *Biological Conservation*, **48**, 41–57.

Pietzarka, U. and Roloff, A. (1993). Waldrandgestaltung unter Berücksichtigung der natürlichen Vegetationsdynamik. *Forstarchiv*, **64**, 107–113.

Pulliam, H.R. (1988). Sources, sinks and population regulation. *American Naturalist*, **132**, 652–661.

Pulliam, H.R. and Danielson, B.J. (1991). Sources, sinks and habitat selection: a landscape perspective on population dynamics. *American Naturalist*, **137** (suppl.), 50–66.

Robertson, P.A. (1992). *Woodland Management for Pheasants*. Forestry Commission Bulletin 106. HMSO, London.

Robertson, P.A., Woodburn, M.I.A. and Hill, D.A. (1993a). Factors affecting winter pheasant density in British woodlands. *Journal of Applied Ecology*, **30**, 459–464.

Robertson, P.A., Woodburn, M.I.A., Neutel, W. and Bealey, C.E. (1993b). Effects of land use on breeding pheasant density. *Journal of Applied Ecology*, **30**, 465–477.

Rosenzweig, M.L. (1991). Habitat selection and population interactions: the search for mechanism. *American Naturalist*, **137** (suppl.), 5–28.

Snow, B. and Snow, D. (1988). *Birds and Berries: A Study of an Ecological Interaction*. Poyser, Calton.

Taylor, P.D., Fahrig, L., Henein, K. and Merriam, G. (1993). Connectivity is a vital element of landscape structure. *Oikos*, **68**, 571–573.

van Dorp, D. and Opdam, P.F.M. (1987). Effects of patch size, isolation and regional abundance on forest bird communities. *Landscape Ecology*, **1**, 59–73.

van Horne, B. (1983). Density as a misleading indicator of habitat quality. *Journal of Wildlife Management*, **47**, 893–901.

Verboom, J., Schotman, A., Opdam, P. and Metz, J.A.J. (1991). European nuthatch metapopulations in a fragmented agricultural landscape. *Oikos*, **61**, 149–156.

Verstrael, T. (1989). Vijf jaar Broedvogel Monitoring Projeckt in Flevoland. *Grauwe Gans*, **5**, 65–92.

Vickery, P.D., Hunter, M.L. and Wells, J.V. (1992). Is density an indicator of breeding success? *Auk*, **109**, 706–710.

Voous, K.H. (1977). *List of Recent Holarctic Bird Species*. British Ornithologists' Union, London.

Woolhouse, M.E.J. (1983). The theory and practice of the species-area effect, applied to the breeding birds of British woods. *Biological Conservation*, **27**, 315–332.

Woolhouse, M.E.J. (1987). On species richness and nature reserve design: an empirical study of UK woodland avifauna. *Biological Conservation*, **40**, 167–178.

APPENDIX. **Scientific names of birds mentioned in the text, following K.H. Voous (1977).**

Abbreviations are given in parentheses for species included in Figure 11.1.

Pheasant *Phasianus colchicus*
Woodpigeon (WP) *Columba palumbus*

Turtle dove *Streptopelia turtur*
Nightjar *Caprimulgus europaeus*
Great spotted woodpecker (GS) *Dendrocopos major*
Lesser spotted woodpecker *Dendrocopos minor*
Tree pipit *Anthus trivialis*
Wren (WR) *Troglodytes troglodytes*
Dunnock (D) *Prunella modularis*
Robin (R) *Erithacus rubecula*
Nightingale *Luscinia megarhynchos*
Blackbird (B) *Turdus merula*
Song thrush (ST) *Turdus philomelos*
Mistle thrush (M) *Turdus viscivorus*
Lesser whitethroat *Sylvia curruca*
Whitethroat (WH) *Sylvia communis*
Garden warbler (GW) *Sylvia borin*
Blackcap (BC) *Sylvia atricapilla*
Chiffchaff (CC) *Phylloscopus collybita*
Willow warbler *Phylloscopus trochilus*
Goldcrest (GC) *Regulus regulus*
Spotted flycatcher (SF) *Muscicapa striata*
Long-tailed tit (LT) *Aegithalos caudatus*
Marsh tit (MT) *Parus palustris*
Willow tit (WT) *Parus montanus*
Coal tit (CT) *Parus ater*
Blue tit (BT) *Parus caeruleus*
Great tit (GT) *Parus major*
Nuthatch (NH) *Sitta europaea*
Treecreeper (TC) *Certhia familiaris*
Jay (J) *Garrulus glandarius*
Starling (SG) *Sturnus vulgaris*
Chaffinch (CH) *Fringilla coelebs*
Greenfinch (GR) *Carduelis chloris*
Linnet (LI) *Carduelis cannabina*
Bullfinch (BF) *Pyrrhula pyrrhula*
Hawfinch *Coccothraustes coccothraustes*
Yellowhammer (Y) *Emberiza citrinella*

12 Population Dynamics of Small Mammals in New Woodlands

J. R. FLOWERDEW and R. C. TROUT

INTRODUCTION

Broadleaved and coniferous woodlands provide favourable habitats for most small rodents and insectivores, although the age and type of woodland may influence the composition of the small mammal community and the abundance of each species (Kirkland and Griffin, 1974; M'Closkey, 1975; Gurnell, 1985; Corbet and Harris, 1991). New woodlands may arise from a variety of management decisions (Buckley and Knight, 1989) but their faunas and floras are likely to parallel the natural woodland succession.

Small mammal distribution and density in woodland depends greatly on its structure and management (Venables and Venables, 1965; Birkan, 1968; M'Closkey, 1975; Ferns, 1979a,b; Tubbs, 1986; Wolk and Wolk, 1982; Aulak, 1970; Gurnell, 1985; Montgomery, 1985) and these in turn determine the quantity and species composition of the ground and field vegetation providing much of the food and cover (Hansson, 1977; Miller and Getz, 1977; Gurnell, 1985). These variables combine as successional influences on the community (Figure 12.1). In addition, the availability and abundance of suitable food sources vary seasonally and annually, with perhaps the most influential being the mast crop (Flowerdew, 1985; Montgomery 1989a; Wilson et al., 1993). The current understanding of the influences of succession and masting on small mammal communities in new woodlands must derive, in the main part, from our knowledge of established woodlands and this forms the background for the present review.

The occurrence of small mammals in woodlands in mainland Britain varies greatly (Table 12.1), according to habitat preferences (Corbet and Harris, 1991). The field vole (*Microtus agrestis*) and harvest mouse (*Micromys minutus*) may make up a large proportion of the community in a grass and herb dominated early-successional stage of woodland, whereas the yellow-necked mouse (*Apodemus flavicollis*) and the common dormouse (*Muscardinus avellanarius*) favour mature deciduous habitats (Harris, 1979; Hurrell and McIntosh, 1984; Montgomery, 1985, 1991; Bright and Morris, 1992).

The Ecology of Woodland Creation. Edited by Richard Ferris-Kaan.

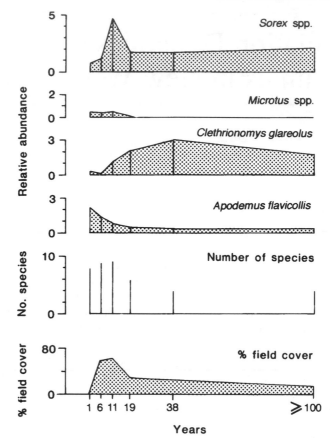

Figure 12.1. Changes in small mammal community structure according to forest age. Data from autumn trapping samples caught over 3 years. From J. Gurnell (1985); modified from E. Wolk and K. Wolk (1982), *Acta Theriologica*, **27**, 45–59 and reproduced by permission from the Polish Academy of Sciences

Assembly rules

Modern ecological theory suggests that each community or assemblage of similar species at a particular successional stage is constrained by 'assembly rules' (Lawton, 1987). For small mammals the rules appear to work by affecting the invasion success, which is a function of the species' functional group; these groups may be defined as herbivore, insectivore or granivore-omnivore, or by body size (Dickman, 1988; Drake, 1990; Fox and Kirkland, 1992). The assembly rule is simply that there is a higher probability that a species entering a community, or species-assemblage, will be drawn from a different functional group until each group is represented (Fox, 1987). The rule is resource based, and increases in resources may produce population increases of one species and/or increases in species richness; for habitats with sufficient resources, the rule may repeat with a second tier of species from each functional group. Tests of these ideas on marsupials, insectivores and rodents add weight to the theory (Fox, 1987; Fox and Kirkland, 1992).

Table 12.1. Occurrence of British mainland non-commensal small mammal species in young and mature successional stages of broadleaved and coniferous woodlands within their distributional ranges

Species		Coniferous		Broadleaved	
		Young	Mature	Young	Mature
Rodents					
Apodemus sylvaticus	Wood mouse	*	*	*	*
A. flavicollis	Yellow-necked mouse	?	?	*	*
Micromys minutus	Harvest mouse	*	?	*	?
Clethrionomys glareolus	Bank vole	*	*	*	*
Microtus agrestis	Field vole	*	?	*	?
Muscardinus avellanarius	Dormouse	?	?	*	*
Insectivores					
Sorex araneus	Common shrew	*	*	*	*
S. minutus	Pygmy shrew	*	*	*	*
Neomys fodiens	Water shrew	?	?	?	?
Talpa europaea	Mole	*	*	*	*

Key: * common: ? occasional or in edge habitats.

This chapter is necessarily concerned with generalisations from a limited number of case studies, and two statements of caution are necessary:

● The most preferred habitats of a small mammal may not support the highest density because they may also be preferred by competing or predatory species (Hansson, 1977).
● Differences in habitat preference may occur within small mammal species both through their geographical range and at their range boundaries (Gurnell, 1985)

In the sections that follow the changes in small mammal community structure and population density expected in early woodland successions in Europe are described, emphasising where possible the parallels with new woodlands in North America. The population dynamics and the influences of masting on the main British species are summarised.

WOODLAND SMALL MAMMAL COMMUNITIES AND WOODLAND SUCCESSION

Geographical variation in small mammal communities

Habitat succession can only influence the small mammal community which is geographically available. European small mammal communities in deciduous and coniferous woodlands vary greatly in their species composition (Table 12.2), partly because of the difficulties of post-glacial invasion to Britain and Ireland (Yalden, 1991). Within these communities usually one or two, and at the most three, species of small mammal occur in each of the three trophic groups within the food spectrum (Gurnell, 1985), and thus feeding habits appear to play at least some part in the structuring of forest small mammal communities (c.f. the 'assembly rules' outlined earlier).

Table 12.2. Examples of small mammal woodland communities in Europe. Modified after J. Gurnell, 1985, *Symposium of the Zoological Society of London*, No. 55, 377–411, and reproduced by permission from Academic Press Ltd, with additional information for North Wales from D. Brown (unpublished)

		Country/habitat						
		Deciduous/ coniferous		Deciduous			Coniferous	
			N.	S.			Poland	
		Ireland	Wales	England	Sweden	Poland	1	> 37 year
g/i	Wood mouse	*	*	*	*	*	*	
g/i	Yellow-necked mouse			O	*	*	*	*
g/i	Other *Apodemus*					*		
g/i	Common dormouse			O		O		
g/i	Harvest mouse			?			*	
f/g	Bank vole	O	*	*	*	*	*	*
f	Field vole		?	?	?	?	*	
f	Other *Microtus*						*	
f	Pine vole *Pitymys*					*		
i	Common shrew		*	*	*	*	*	*
i	Pygmy shrew	*	?	*	*	*	*	*
i	Water shrew		?	?	?			
i	Mole		+	+	+	+	+	+

Status: * present, ? occasional, O present but local distribution, + assumed present.
Diet: g: granivore/omnivore, i: insectivore, f: folivore (herbivore).

Changes in small mammal communities with age/successional stage

It is clear from Tables 12.1 and 12.2 that the small mammal species composition of coniferous and deciduous woodland in the same geographical area is often similar, differing by only a few species, if at all. A parallel situation is described from studies in North American woodlands (Kirkland, 1977); here it was found that the small mammal species richness (mean number of species per sampling plot) was greater in mature diverse broadleaved forest than in mature coniferous plots (mainly red spruce (*Picea rubens*)), but the difference was, on average, only 1.25 species per plot. It was also found that rodent species-richness declined in young (0–5 years) broadleaved successions, increasing later (6–15 years) and that the opposite was true for coniferous plots. The greater vegetational diversity of the older deciduous areas was thought to account for this. The shrews (*Sorex* spp.) showed less variation in species-richness with time. The relative abundance (catch per unit sampling effort) of all the small mammal species in both mature (25 years old) and sapling/young pole tree stages (6–15 years old) of coniferous forest and their relative abundances in similar stages of the deciduous forest were strongly correlated.

Competition may also affect community composition (Gurnell, 1985; Montgomery, 1989a). One species may dominate another, causing low densities or its absence, and the influence of one species on another may be seasonal. The diversity of the habitat in structure and species composition may also allow normally competing species to coexist.

The natural successional changes initiating British woodland will often start with a grassland community being invaded by, for example, beech (*Fagus sylvatica*) (Tansley, 1939) or Scots pine (*Pinus sylvestris*) (Crompton and Sheail, 1975). In managed situations, however, clearfelled (possibly burned) woodland, rough ground and previously cultivated land are often sites for the establishment of new woodlands. Many of these habitats contain bare soil and agricultural weeds are likely to form the initial plant community (Davies, 1987; Williamson, 1992; Rodwell and Patterson, Chapter 5). Unless management intervenes, the small mammal community will change gradually, as the grassland field layer develops and then declines as the canopy grows, following classical successional stages (Cousens, 1974). The nature of the field layer will vary with soil conditions, the shade cast by the growing trees and with the influence of the increasing leaf litter. At the mature woodland stage the canopy cover may decline as some trees die through disease or are blown down or removed during management, allowing a resurgence of field layer growth.

The diversity and number of plant species in the ground and field layer is expected to increase in the pioneer phase and then decrease as the tree canopy shades out the ground and field vegetation (Kirkland, 1977; Wolk and Wolk, 1982; Gurnell, 1985). During a forest succession there are parallel changes in the small mammal and plant communities (Figure 12.1). These long-term changes, particularly the loss of *Microtus* species with the deterioration of their grassland habitat, are common to broadleaved and coniferous habitats in Europe and North America (Birkan, 1968; Gashwiler, 1979; Grant, 1971; Kirkland, 1977; Ferns, 1979a,b; Charles, 1981; Wolk and Wolk, 1982; Gurnell, 1985; Richards, 1985; Montgomery, 1985).

The influence of ground vegetation biomass on field vole density has been shown experimentally (MacVicar and Trout, 1994). In new woodland plantations in plots of previously ploughed arable land, field vole density was greatest in both spring and autumn where ground vegetation biomass was greatest (Figure 12.2); in addition, the

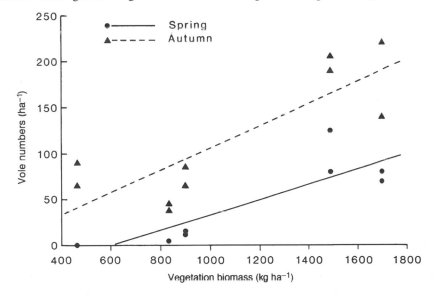

Figure 12.2. Relationships between field vole (*Microtus agrestis*) density and ground vegetation biomass in spring (●) and autumn (▲), on plots of new deciduous woodland planted in previously arable land. From H.J. MacVicar and R.C. Trout (1994), reproduced by permission of The British Grassland Society

reduction of biomass by mowing three times each year in formerly vole-rich plantations significantly reduced signs of field voles and damage to small trees by them (Figure 12.3). Similar vegetational influences occur in North American forests where herbaceous growth is retarded after clearfelling, slowing down the small mammal succession (Gashweiler, 1970).

Table 12.3. Small mammal species composition in new woodland and later succession. (a) New deciduous woodland on ex-agricultural fields in southern England with no tree canopy (data from R.C. Trout and H.J. MacVicar, unpublished). (b) Corsican pine (*Pinus nigra*) plantation at Newborough, Warren, Anglesey, previously sand dunes with marram (*Ammophila arenaria*), planted in 1949. Some thick undergrowth with marram and bare needle carpet in 1960 (trees *c.* 3 m high) changed to little ground vegetation under closed canopy, with brashings removed, by 1964 (trees *c.* 5 m high). Data from L.S.V. Venables and U.M. Venables, 1965, *Nature in Wales*, **9**, 171–184

	Percentage of captures		
	(a) New woodland	(b) Corsican pine plantation	
Species		March 1960–March 1962	October 1962–October 1964
	3-year period		
Field vole	30	51	9
Bank vole	< 1	9	35
Wood mouse	33	18	16
Common shrew	24	21	40
Pygmy shrew[a]	8		
Harvest mouse	5	0	0
Total captures	1642	33	55

[a] Pygmy shrew, mole and water shrew were also recorded in the plantation.

The pioneer small mammal species in Britain are the wood mouse (*Apodemus sylvaticus*), which even occupies bare soil in ploughed fields (Johnson *et al.*, 1992; Tew *et al.*, 1992), and, to a lesser extent, the common shrew (*Sorex araneus*) and pygmy shrew (*Sorex minutus*). In clearfelled woodland and other poorly vegetated communities, it is likely that these shrews and possibly the water shrew (*Neomys fodiens*) will invade the wood-land within the first few years (Table 12.3(a)), as soon as some field layer vegetation is present (Venables and Venables, 1965; Hansson, 1971; Churchfield, 1980). Later in the succession, bank voles (*Clethrionomys glareolus*) will invade as *Microtus agrestis* decrease in density (Table 12.3(b)).

The effects of coppicing on successional changes

Coppicing as a management practice is common in established deciduous woodland and may become an important management tool in new woodlands, especially farm wood-lands. It consists of the felling of the small trees and bushes, the coppice, on a regular cycle, removing the canopy layer and often leaving isolated 'standard' trees behind, usually oaks (Rackham, 1975).

The coppice succession has a marked effect on the small mammal community (Gurnell *et al.*, 1992). During coppicing the soil is disturbed, leaving little ground cover, but by the autumn of the third year 70% of the ground is covered, reaching almost 100% by year 10. The height of the field layer vegetation reaches more than 2 m in the first year but covers only a small percentage of the ground. By 10 years this declines to about 1 m over most of the ground. Coppicing allows two rodent species, yellow-necked mouse and harvest mouse, not usually abundant in woodland, to enter the community (Table 12.4). Later, the community stabilises with fewer species more characteristic of mature woodland. Thirty year old coppice has a smaller proportion of bank voles than mature oak woodland (Table 12.4), because large gaps appear in the canopy due to woodland senescence; the additional light promotes the growth of dense vegetation such as bracken (*Pteridium aquilinum*) and bramble (*Rubus fruticosus*), a habitat favoured by bank voles (Southern and Lowe, 1968).

Table 12.4. Relative abundance of species (% total numbers) in five ages of coppice at Bradfield Woods, Suffolk, and in an oak (*Quercus robur*) woodland in southern England. After J. Gurnell *et al.*, 1992, in *Ecology and Management of Coppice Woodlands*, G.P. Buckley (ed.), and reproduced by permission from Chapman and Hall, London

Species		Age of coppice (years)					
		1	3	10	20	30	Oak
Microtus agrestis	Field vole	3	2	< 1	–	–	< 1
Clethrionomys glareolus	Bank vole	10	37	19	16	10	41
Apodemus sylvaticus	Wood mouse	40	18	52	55	59	41
A. flavicollis	Yellow-necked mouse	2	2	10	11	14	4
Micromys minutus	Harvest mouse	20	5	< 1	–	< 1	< 1
Sorex araneus	Common shrew	23	25	17	17	14	13
S. minutus	Pygmy shrew	< 1	7	2	2	2	2
Neomys fodiens	Water shrew	< 1	< 2	< 1	–	–	< 1
Number of species >1%		6	8	4	5	5	5

Note that trapping methods (live traps) may under-represent pygmy shrews.

In sweet chestnut (*Castanea sativa*) coppice of 2–15 years, a trend for *Apodemus sylvaticus* numbers to increase with age of coppice has been found (Toms, 1990). *C. glareolus* was absent from 6 month sweet chestnut coppice, abundant in year 2 and less abundant or absent in the later stages.

General effects of successional changes on small mammals

In the first year after planting trees, small mammal communities are likely to be depauperate if there is little grassland; only the wood mouse and possibly common and pygmy shrews may be present. With the growth of the field layer a diverse grassland small mammal community appears, usually dominated by field voles and with fewer wood mice, possibly three shrew species and bank voles and also having harvest mice, if present in the locality. The development of this community may be rapid in new woodlands planted on nutrient-rich agricultural land where a luxuriant herb and field layer develops. Later, as the grassland gives way to woodland herbs, and litter accumulates under the closed canopy, the field voles and harvest mice disappear (except from small areas of favourable habitat such as woodland edge and glades under breaks in the canopy) and wood mice dominate the community with bank voles and common and pygmy shrews. At this time, yellow-necked mice and dormice may occur within their distributional ranges.

Each woodland will have local factors influencing the pattern of colonisation, habitat exploitation and community succession. In addition, the pattern of population dynamics in each species will modify these communities on a seasonal basis so that one species may not be dominant throughout the year. Some of the factors peculiar to woodland which affect small mammals on a long-term and a seasonal basis are discussed below.

MASTING AND OTHER ECOLOGICAL FACTORS AFFECTING SMALL MAMMAL FOOD SUPPLIES

After maturity, many broadleaved tree species produce heavy crops of fruits and seeds synchronously with a characteristic periodicity (Fowells, 1965; Grubb, 1977; Silvertown, 1980). This is known as 'masting' and even relatively young trees of some species (e.g. ash (*Fraxinus excelsior*)) will still show periodic heavy fruiting.

Broadleaved species noted for masting and which have recorded effects on European rodent populations are sessile oak (*Quercus robur*) and pedunculate oak (*Quercus petraea*) (Smyth, 1966; Watts, 1969; Gurnell, 1981; Smal and Fairley, 1982; Pucek *et al.*, 1993; Mallorie and Flowerdew, 1994), ash (Flowerdew and Gardner, 1978; Flowerdew, 1993), beech (Bobek, 1969; Hansson, 1971; Jensen, 1982; Le Louarn and Schmitt, 1972), hornbeam (*Carpinus betulus*) and Norway maple (*Acer platanoides*) (Pucek *et al.*, 1993). Data on fruiting and seed production for these and many other species are summarised by Grodzinski and Sawicka-Kapusta (1970) and Silvertown (1980). Many coniferous species also show masting (Silvertown, 1980; Pucek *et al.*, 1993) and significant effects of seed production on small rodent populations may also occur. In North American Douglas fir (*Pseudotsuga menziesii*) forests seed abundance affects reproduction and numbers of deer mice (*Peromyscus maniculatus*) (Gashweiler, 1979). Food supply is an important limiting factor for small mammals and masting generally influences small rodent populations by improving survival and allowing higher densities than 'normal' as well as lengthening the breeding season (Flowerdew, 1985). There may be further effects of masting on the woodland invertebrate population and consequently on the shrews (and rodents) but evidence of this is lacking.

The periodicity, timing and size of the mast crop varies with species. In the oaks and beech, the crop is very heavy (production of $50+ \mathrm{g\,m^{-2}}$) and in the case of beech may be relatively frequent (Hilton and Packham, 1986). The oaks, however, produce heavy crops of mast (up to $64.5 \mathrm{g\,m^{-2}}$) once every 6–7 years and only moderate crops every 3–4 years (Jones, 1959). Ash produces smaller weights of fruit (containing up to $18 \mathrm{g\,m^{-2}}$ of edible seed), more or less biennially but there are some exceptions as moderately heavy crops may be produced two years in succession (Flowerdew and Gardner, 1978; Flowerdew, unpublished). Peak ash fruit fall is usually recorded in mid-winter (Gardner, 1977) whereas peak acorn and beech mast fall is during October (Gurnell, 1981; Le Louarn and Schmitt, 1972).

Small rodent food supplies will also be affected by other seed predators; the absence of grey squirrels (*Sciurus carolinensis*) may allow seeds to be available on the woodland floor in greater numbers and for longer periods than when they are present (Ashby, 1967; Gurnell, 1981).

SMALL MAMMAL POPULATION DYNAMICS IN WOODLAND

The dynamics of many small mammal species are known to a greater or lesser extent from detailed studies in a wide variety of habitats (Flowerdew, 1985; Tamarin, 1985; Montgomery, 1989a; Corbet and Harris, 1991). Population studies from early successional grasslands, young plantations and old-field habitats are likely to be similar to many new woodland situations, whereas studies from mature woodland may be relevant only to older stages of new woodland development where there is little management of the ground vegetation. The following descriptions of small mammal population dynamics give an outline of the intra- and inter-annual fluctuations to be expected in woodlands and the influence of masting on these dynamics. The seasonal and multiannual fluctuations in numbers found in each species will have profound effects on the relative proportions of each species found in the woodland community.

The field vole

Field voles commonly reach 50 ha^{-1} or more in grassland (Chitty et al., 1968; Tapper, 1976; Charles, 1981; Richards, 1985; Gipps and Alibhai, 1991). They also fluctuate within years, often with spring lows and autumn peaks, densities are generally highest in young woodlands (Birkan, 1968; Southern, 1970; Gurnell, 1981; Gurnell et al., 1992), although they may be absent from 15–25 year plantations (Montgomery, 1985). Multiannual fluctuations also occur; in young conifer plantations and permanent grasslands such as those in northern England and Scotland, short-tailed voles show regular (3–4 years) cyclical fluctuations with considerable synchrony over large geographical areas (Chitty, 1952; Charles, 1981; Taylor et al., 1988).

Following heavy masting in mature deciduous woodlands, field voles reach higher densities than usual and show winter breeding (Smyth, 1966; Gurnell, 1981). Given that they are graminivorous, these associations are difficult to explain and the immediate cause of such breeding is unknown.

When new plantings take place on fertile agricultural land (such as with the Farm Woodland Scheme and Farm Woodland Premium Scheme in the UK), vole numbers respond to the luxuriant vegetation between the rows of trees and can cause considerable tree losses (Figure 12.2). However, the voles can be almost eliminated by repeated mowing (Figure 12.3).

The bank vole

Bank voles are common in young and mature deciduous and coniferous woodland (Alibhai and Gipps, 1991) but they may suffer from competition from M. agrestis (Flowerdew et al., 1977; Gurnell, 1985). Populations in mature woodlands show a strong preference for habitats with good ground cover (Southern and Lowe, 1968); this is a major factor influencing population density and will also apply to new woodlands. Population fluctuations usually follow a seasonal pattern with an autumn peak and a

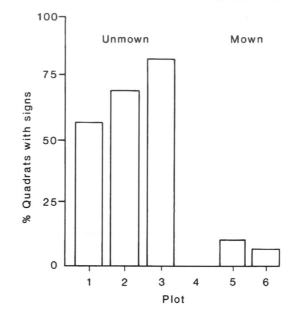

Figure 12.3. Results of experimental mowing (three times per year) on signs of field vole (*Microtus agrestis*) abundance (% of quadrats with faeces or grass clippings present) in new deciduous woodland. Prior to the experiment, 70.8% of quadrats within plots showed signs of *M. agrestis*. From MacVicar and Trout, unpublished

spring trough; less commonly they increase in spring or remain stable (Alibhai and Gipps, 1985; Mallorie and Flowerdew, 1994). This annual pattern may follow the dynamics of wood mice very closely (Gurnell, 1981) or be independent of them (Mallorie, 1992). Cyclical (multi-annual) fluctuations may be present in more northerly populations but are probably not common in Britain (Southern and Lowe, 1982; Alibhai and Gipps, 1985).

Bank voles show strong reactions to heavy mast crops, with overwinter breeding, or breeding later than usual (oaks: Smyth, 1966; beech: Louarn and Schmitt, 1972; Bobek, 1969; Jensen, 1982). After 'mast years' in such woodlands autumn numbers were higher than in non-mast years, providing the habitats had sheltering ground vegetation (Hansson, 1971). Densities are significantly higher (by a factor of 2) in the summer following a good or heavy seed crop than after a poor one, and the rates of population change from winter to summer are also significantly higher (Mallorie and Flowerdew, 1994). There are also delayed reactions to heavy ash fruit fall, where numbers increase from summer to winter following the winter of the mast crop (Flowerdew, unpublished); this effect has also been shown in Danish beech woods by Jensen (1982).

In coniferous woodlands, the seed fall is often delayed until the year following flowering, or longer, and the amount is often less than from broadleaved trees (Turcek, 1967). Thus, irruptions of bank voles (e.g. 137 ha^{-1} (Newson, 1963)) and other seed-eating species such as wood mice are more likely in broadleaved woodland. High densities often lead to reduced reproduction in these species (Alibhai and Gipps, 1985; Montgomery, 1989a,b) and thus population declines may occur at unusual times of the year.

The wood mouse

Wood mice are usually present in woodland throughout the succession (Flowerdew, 1991) and should form a major component of any new woodland small mammal community. In broadleaved woodland, numbers are often 1–40 ha^{-1}, but may be 130–200 ha^{-1} following masting (Gurnell, 1981). Coniferous woodlands probably support fewer wood mice than broadleaved woodlands (Gurnell, 1981; Venables and Venables, 1965; Montgomery, 1985; Mallorie, 1992).

The population dynamics of wood mice are well understood (Flowerdew, 1985; 1991; Wilson et al., 1993). In deciduous woodland there are regular annual fluctuations with autumn/early winter peaks, declines in spring and autumn increases, often preceded by a period of relative stability in summer. In mast years peaks last longer and are later (Flowerdew, 1985; Montgomery, 1989a); a constant high density or a continuous increase from winter to winter may occur occasionally. In the typical fluctuation the increase into the winter is density dependent and this usually limits winter numbers. However, after heavy mast crops density dependence may be overridden and, after high peaks, the decline to the spring is often not very severe (Gurnell, 1981; Flowerdew, 1985; Montgomery, 1989b).

The influence of masting is generally immediate, causing high winter peaks and good overwinter survival, often with an extension of breeding into the winter (Smal and Fairley 1982; Flowerdew, 1985; Gurnell, 1985; Tubbs, 1986; Mallorie and Flowerdew, 1994). Winter breeding after heavy masting has been recorded from deciduous woodlands (oaks: Smyth, 1966; D. Brown unpublished; beech: Louarn and Schmitt, 1972) and immigration is high under similar circumstances (Hansson, 1971; Flowerdew, 1976).

The yellow-necked mouse

Yellow-necked mice are mainly inhabitants of mature deciduous woodland in Britain and they are restricted to the south and south-east and the Welsh borders (Montgomery, 1991). They prefer areas with little ground cover but with dense cover between 1 and 5 m (Montgomery, 1991); areas with high energy fruits such as yew (*Taxus baccata*) appear to be especially favourable (Montgomery, 1985).

Population fluctuations are often similar to those of wood mice, although they may be at low density with scattered centres of population dispersed throughout the woodland (Gurnell, 1985; Montgomery, 1985). Generally the autumn increase is later than for *A. sylvaticus* and the spring decline less precipitate. Yellow-necked mice respond to heavy masting with high peak numbers and better spring breeding (Gurnell, 1985; Montgomery, 1989a, 1991; Pucek et al., 1993). In Britain, even in the vicinity of high populations in deciduous woodlands, densities are likely to be low in grassland and conifer plantations and only moderate in secondary deciduous woodland (Montgomery, 1985), and thus densities in new woodlands are likely to be low, if they are present at all.

The harvest mouse

Harvest mice are found in tall vegetation such as grassland, reed beds and woodland edge (Harris and Trout, 1991), as well as in early coppice (see earlier section). Their distribution spreads throughout southern England, but in Wales and in northern Britain as far as

Edinburgh, the populations are isolated and local. Numbers may reach up to 250 ha^{-1} in autumn and early winter (Trout, 1978). In young plantations harvest mice are found mainly from late summer to winter with an autumn peak and they are likely to appear in any new woodland with sufficient tall ground vegetation. The effects of masting on this species are unknown.

The common dormouse

Common dormice are widely but patchily distributed in Britain from mid-Wales, the Midlands and Suffolk southwards (Morris, 1991). Deciduous woodland with plenty of shrub growth and scrub such as bramble is the major habitat, although they are likely to be absent before the canopy of young woodland closes. Heavily fruiting trees and shrubs such as *Fagus sylvatica, Quercus* spp. *Castanea sylvatica*, wayfaring tree (*Viburnum lantana*), blackthorn (*Prunus spinosa*) and hazel (*Corylus avellana*) and the presence of arboreal runways provided by honeysuckle (*Lonicera peryclimenum*) growth are favoured (Morris, 1991; Bright and Morris, 1992). However, suitable habitats also include damp woods with plenty of mosses (Morris, 1991) and the edges of conifer plantations as long as they include deciduous shrubs (Hurrell and McIntosh, 1984). Short-rotation (7–10 years) coppice is unfavourable for dormice while longer rotations (15–20 years), which have a much greater development of shrubs and food production (flowers, nuts and fruits), are preferred (Bright and Morris, 1989, 1992). Because their distribution is so patchy it will probably be unusual for dormice to colonise new woodlands, but if suitable habitats are available then it should be possible.

Dormice reach only 8–10 ha^{-1} even in prime habitat, but in continental Europe they may increase sufficiently in numbers after a heavy mast crop to cause damage to young trees (Morris, 1991). Numbers probably increase during the breeding season (May–September), reaching an autumn peak, then decline over winter with some mortality occurring during hibernation because of insufficient fat reserves and/or predation (Morris, 1991).

Common, pygmy and water shrews

The shrews (*Sorex* spp. and *Neomys fodiens*), as insectivores, are not affected by masting, but they form a major component of the small mammal community in most woodlands. The common shrew is widespread throughout England, Wales and Scotland but is absent from Ireland and many islands (Churchfield, 1991a). In 18-year-old larch/spruce woodland a peak of 33 ha^{-1} was recorded (Dickman, 1980), similar to the density of 24–28 ha^{-1} found in deciduous woodland (Shillito, 1960). In rough grassland and young successional stages of woodland 42–67 ha^{-1} may be found (Yalden, 1974), although lower densities are also possible (Pernetta, 1977). Numbers follow a regular annual pattern, peaking in late summer with many juveniles present and then declining to the next spring probably with reduced activity above ground (Churchfield, 1991a).

Pygmy shrews are widespread in the British Isles and Ireland (Churchfield, 1991b). They are common in habitats which have plenty of ground cover, but are usually less abundant than common shrews. However, high densities have been found in 'bare' mature Sitka spruce plantations in Ireland (Grainger and Fairley, 1978). Annual fluctuations are similar to those of common shrews (Churchfield, 1991b). In old-field grassland with shrubs densities reached 11–12 ha^{-1} (Pernetta, 1977), and they may be present at low densities in mature broadleaved and coniferous woodland (Venables and Venables, 1965; Montgomery, 1985). The species is, however, often absent from coniferous plantations (Dickman, 1980; Don, 1979).

The water shrew is also widespread in Britain, absent from Ireland and is only locally distributed in Scotland (Churchfield, 1991c). It will enter grassland and woodland well away from water. It is probably only an occasional inhabitant of most woodlands and may be seasonally common (Ferns, 1979a; Churchfield, 1991c).

CONCLUSIONS

New woodland will first affect the small mammal community by the changes in ground flora associated with natural or managed succession. The rodent densities and species composition may be influenced to a greater extent by successional changes in the ground vegetation and by masting than the insectivorous shrews. Community composition is dependent on the small mammal species which are geographically available, and the species number is likely to increase in the early years, stabilising at a lower level with woodland maturity. Further changes to the community (alterations in species number or relative density) may arise following management (e.g. coppicing or weed control before canopy closure) or later with the senescence or felling of the wood. In addition, the normal seasonal fluctuations in numbers of small mammals are likely to alter the relative abundance of the species within the community. A dominant factor affecting rodent population density and fluctuations as the wood develops is likely to be the masting characteristic(s) of the tree species present. Heavy masting will affect rodent numbers immediately, with increased winter densities and survival, and delayed effects are also possible; winter breeding may also be stimulated. The intensity of the effects of masting vary between the rodent species.

New woodland is likely to be beneficial to small mammal populations as it has the potential, given sympathetic management, to provide a greater diversity of habitat and hence small mammals than agricultural crops, pasture or derelict urban ground. In addition, the small mammal community will develop with the ecological succession in the woodland. Early small mammal communities in new woodland planted in previously fertile agricultural land are likely to be species-rich and at a relatively high density in the luxuriant ground vegetation. This contrasts with the meagre ground vegetation found in many set-aside fields where the mowing regime prevents the development of a diverse and complex field layer and leads to a depauperate small mammal community.

ACKNOWLEDGEMENTS

R. C. Trout is funded by the Land Use, Conservation and Countryside Group, MAFF.

REFERENCES

Alibhai, S.K. and Gipps, J.H.W. (1985). The population ecology of bank voles. *Symposium of the Zoological Society of London*, **55**, 277–305.

Alibhai, S.K. and Gipps, J.H.W. (1991). Bank vole *Clethrionomys glareolus*. In: Corbet, G.B. and Harris, S. (eds). *The Handbook of British Mammals*. Blackwell Scientific Publications, Oxford, 192–203.

Ashby, K.R. (1967). Studies on the ecology of field mice and voles (*Apodemus sylvaticus, Clethrionomys glareolus* and *Microtus agrestis*) in Houghall Wood, Durham. *Journal of Zoology, London*, **152**, 389–513.

Aulak, W. (1970). Small mammal communities in the Bialowieza National Park. *Acta Theriologica*, **15**, 465–515.

Birkan, M. (1968). Repartition ecologique et dynamique des populations d'*Apodemus sylvaticus* et *Clethrionomys glareolus* en pinede a Rambouillet. *Terre et Vie*, **3**, 231–273.

Bobek, B. (1969). Survival, turnover and production of small rodents in a beech forest. *Acta Theriologica*, **14**, 191–210.

Bright, P. and Morris, P.A. (1989). *A Practical Guide to Dormouse Conservation*. Occasional publication of The Mammal Society No. 11. The Mammal Society, London.

Bright, P. and Morris, P.A. (1992). *Dormice*. The Mammal Society, London.

Buckley, G.P. and Knight, D.G. (1989). The feasibility of woodland reconstruction. In: Buckley, G.P. (ed.). *Biological Habitat Reconstruction*. Belhaven Press, London, 171–188.

Charles, W.N. (1981). Abundance of field voles (*Microtus agrestis*) in conifer plantations. In: Last, F.T. and Gardiner A.S. (eds). *Forest and Woodland Ecology*. ITE/NERC, Cambridge, 135–137.

Chitty, D. (1952). Mortality among voles (*Microtus agrestis*) at Lake Vernwy, Montgomeryshire, in 1936–39. *Philosophical Transactions of the Royal Society (B)*, **236**, 505–552.

Chitty, D., Pimentel, D. and Krebs, C.J. (1968). Food supply of overwintered voles. *Journal of Animal Ecology*, **37**, 113–120.

Churchfield, S. (1980). Population dynamics and the seasonal fluctuations in numbers of the common shrew in Britain. *Acta Theriologica*, **25**, 415–424.

Churchfield, S. (1991a). Common shrew *Sorex araneus*. In: Corbet, G.B. and Harris, S. (eds). *The Handbook of British Mammals*. Blackwell Scientific Publications, Oxford, 51–58.

Churchfield, S. (1991b). Pygmy shrew *Sorex minutus*. In: Corbet, G.B. and Harris, S. (eds). *The Handbook of British Mammals*. Blackwell Scientific Publications, Oxford, 60–64.

Churchfield, S. (1991c). Water shrew *Neomys fodiens*. In: Corbet, G.B. and Harris, S. (eds). *The Handbook of British Mammals*. Blackwell Scientific Publications, Oxford, 64–68.

Corbet, G.B. and Harris, S. (eds). (1991). *The Handbook of British Mammals*. Blackwell Scientific Publications, Oxford.

Cousens, J. (1974). *An Introduction to Woodland Ecology*. Oliver and Boyd, Edinburgh.

Crompton, G. and Sheail, J. (1975). The historical ecology of Lakenheath Warren in Suffolk, England: a case study. *Biological Conservation*, **8**, 299–313.

Davies, R.J. (1987). *Trees and Weeds – Weed Control for Successful Tree Establishment*. Forestry Commission Handbook No. 2. HMSO, London.

Dickman, C.R. (1980). Estimation of population density in the common shrew, *Sorex araneus*, from a conifer plantation. *Journal of Zoology, London*, **192**, 550–552.

Dickman, C.R. (1988). Body size, prey size, and community structure in insectivorous mammals. *Ecology*, **69**, 569–580.

Don, B.A.C. (1979). Gut analysis of small mammals during a sawfly (*Cephalcia lariciphila*) outbreak. *Journal of Zoology, London*, **188**, 290–294.

Drake, J.A. (1990). Communities as assembled structures: do rules govern patterns? *Trends in Ecology and Evolution*, **5**, 159–164.

Ferns, P.N. (1979a). Successional changes in the small mammal community of a young larch plantation in south west Britain. *Mammalia*, **43**, 439–452.

Ferns, P.N. (1979b). Growth, reproduction and residency in a declining population of *Microtus agrestis*. *Journal of Animal Ecology*, **48**, 739–758.

Flowerdew, J.R. (1976). The effect of a local increase in food supply on woodland mice and voles. *Journal of Zoology, London*, **180**, 509–513.

Flowerdew, J.R. (1985). The population dynamics of wood mice and yellow-necked mice. *Symposium of the Zoological Society of London*, No. 55, 315–338.

Flowerdew, J. (1993). *Mice and Voles*. Whittet Books, London.

Flowerdew, J.R. and Gardner, G. (1978). Small rodent populations and food supply in a Derbyshire ashwood. *Journal of Animal Ecology*, **47**, 725–740.

Flowerdew, J.R., Hall, S.J.G. and Brown, J. Clevedon (1977). Small rodents, their habitats and the effects of flooding at Wicken Fen, Cambridgeshire. *Journal of Zoology, London*, **182**, 323–342.

Fowells, H.A (1965). *Sylvics of Forest Trees of the United States*. Agriculture Handbook No. 271, U.S.D.A. Forest Service, Washington DC.

Fox, B.J. (1987). Species assembly and the evolution of community structure. *Evolutionary Ecology*, **1**, 201–213.

Fox, B.J. and Kirkland, G.L. (1992). An assembly rule for functional groups applied to North American soricid communities. *Journal of Mammalogy*, **73**, 481–503.

Gardner, G. (1977). The reproductive capacity of *Fraxinus excelsior* on the Derbyshire limestone. *Journal of Ecology*, **65**, 107–118.

Gashweiler, J.S. (1970). Further study of conifer seed survival in a western Oregon clearcut. *Ecology*, **51**, 849–854.

Gashweiler, J.S. (1979). Deer mouse reproduction and its relationship to the tree seed crop. *American Midland Naturalist*, **102**, 95–104.

Gipps, J.H.W. and Alibhai, S.K. (1991). Field vole *Microtus agrestis*. In: Corbet, G.B. and Harris, S. (eds). *The Handbook of British Mammals*. Blackwell Scientific Publications, Oxford, 203–208.

Grainger, J.P. and Fairley, J.S. (1978). Studies on the biology of the pygmy shrew, *Sorex minutus*, in the west of Ireland. *Journal of Zoology, London*, **186**, 109–141.

Grant, P.R. (1971). The habitat of *Microtus pennsylvanicus* and its relevance to the distribution of this species on islands. *Journal of Mammalogy*, **52**, 351–361.

Grodzinski, W. and Sawicka-Kapusta, K. (1970). Energy values of tree-seeds eaten by small mammals. *Oikos*, **21**, 52–58.

Grubb, P.J. (1977). The maintenance of species-richness in plant communities: the importance of the regeneration niche. *Biological Reviews*, **52**, 107–145.

Gurnell, J. (1981). Woodland rodents and tree seed supplies. In: Chapman, J.A. and Pursley, D. (eds). *The Worldwide Furbearer Conference Proceedings*, Falls Chard, Virginia, USA. R.R. Donnelly and Sons Co., 1191–1214.

Gurnell, J. (1985). Woodland rodent communities. *Symposium of the Zoological Society of London*, No. 55, 377–411.

Gurnell, J., Hicks, M. and Whitbread, S. (1992). The effects of coppice management on small mammal populations. In: Buckley, G.P. (ed.). *Ecology and Management of Coppice Woodlands*. Chapman and Hall, London, 213–232.

Hansson, L. (1971). Small rodent food, feeding and population dynamics. *Oikos*, **22**, 183–198.

Hansson, L. (1977). Spatial dynamics of field voles *Microtus agrestis* in heterogeneous landscapes. *Oikos*, **29**, 539–544.

Harris, S. (1979). History, distribution, status and habitat requirements of the harvest mouse (*Micromys minutus*) in Britain. *Mammal Review*, **9**, 159–171.

Harris, S. and Trout, R.C. (1991). Harvest mouse *Micromys minutus*. In: Corbet, G.B. and Harris, S. (eds) *The Handbook of British Mammals*. Blackwell Scientific Publications, Oxford, 233–239.

Hilton, G.M. and Packham, J.R. (1986). Annual and regional variation in English beech mast (*Fagus sylvatica* L.). *Arboriculture Journal*, **10**, 3–14.

Hurrell, E. and McIntosh, G. (1984). Mammal Society dormouse survey, January 1975 to April 1979. *Mammal Review*, **14**, 1–18.

Jensen, T.S. (1982). Seed production and outbreaks of non-cyclic rodent populations in deciduous forests. *Oecologia*, **54**, 184–192.

Johnson, I.P., Flowerdew, J.R. and Hare, R. (1992). Population and diet of small rodents and shrews in relation to pesticide usage. In: Greig-Smith, P., Frampton, P. and Hardy, T. (eds). *Pesticides, Farming and the Environment*. HMSO, London, 144–156.

Jones, E.W. (1959). *Quercus* L. Biological flora of the British Isles. *Journal of Ecology*, **47**, 169–222.

Kirkland, G.L. Jr. (1977). Responses of small mammals to the clearcutting of northern Appalachian forests. *Journal of Mammalogy*, **58**, 600–609.

Kirkland, G.L. Jr. and Griffin, R.J. (1974). Microdistribution of small mammals at the coniferous-deciduous forest ecotone in northern New York. *Journal of Mammalogy*, **55**, 417–427.

Lawton, J.H. (1987). Are there assembly rules for successional communities? In: Gray, A.J., Crawley, M.J. and Edwards, P.J. (eds). *Colonization, Succession and Stability*. 26th Symposium of the British Ecological Society. Blackwell Scientific Publications, Oxford, 225–244.

Louarn, H. Le and Schmitt, A. (1972). Relations observées entre la production de faines et la dynamique de population du mulot, *Apodemus sylvaticus* L., en forêt de Fontainebleau. *Annales des Sciences Forestieres*, **30**, 205–214.

Mallorie, H.C. (1992). The population dynamics of small rodents in coniferous woodland. PhD thesis, Cambridge University.

Mallorie, H.G. and Flowerdew, J.R. (1994). Woodland small mammal ecology in Britain. A preliminary review of the Mammal Society survey of wood mice (*Apodemus sylvaticus*) and bank voles (*Clethrionomys glareolus*). *Mammal Review*, **24**, 1–15.

M'Closkey, R.T. (1975). Habitat succession and rodent distribution. *Journal of Mammalogy*, **56**, 950–955.

MacVicar, H.J. and Trout, R.C. (1994). Vegetation management to manipulate field vole (*Microtus agrestis*) populations in grassy plantations. In: Haggar, R.J. and Peel, S. (eds). *Grassland Management and Nature Conservation*. British Grassland Society Occasional Symposium No. 28, Leeds, 302–305.

Miller, D.H. and Getz, L.L. (1977). Factors influencing local distribution and species diversity of forest small mammals in New England. *Canadian Journal of Zoology*, **55**, 806–814.

Morris, P.A. (1991). Common dormouse *Muscardinus avellanarius*. In: Corbet, G.B. and Harris, S. (eds). *The Handbook of British Mammals*. Blackwell Scientific Publications, Oxford, 259–264.

Montgomery, W.I. (1985). Interspecific competition and the comparative ecology of two congeneric species of mice. In: Cook, L.M. (ed.) *Case Studies in Population Biology*. Manchester University Press, Manchester, 126–187.

Montgomery, W.I. (1989a). Population regulation in the wood mouse, *Apodemus sylvaticus*. I. Density dependence in the annual cycle of abundance. *Journal of Animal Ecology*, **58**, 465–476.

Montgomery, W.I. (1989b). *Peromyscus* and *Apodemus*: patterns of similarity in ecological equivalents. In: Kirkland, G.L. Jr. and Layne, J.N. (eds). *Advances in the Study of Peromyscus* (Rodentia). Texas Technical University Press, Lubbock, 293–365.

Montgomery, W.I (1991). Yellow-necked mouse *Apodemus flavicollis*. In: Corbet, G.B. and Harris, S. (eds). *The Handbook of British Mammals*. Blackwell Scientific Publications, Oxford, 229–233.

Montgomery, W.I., Wilson, W.L., Hamilton, R. and McCartney, P. (1991). Dispersion in the wood mouse *Apodemus sylvaticus*: variable resources in time and space. *Journal of Animal Ecology*, **60**, 179–192.

Newson, R. (1963). Differences in numbers, reproduction and survival between two neighbouring populations of bank voles (*Clethrionomys glareolus*). *Ecology*, **44**, 110–120.

Pernetta, J. (1977). Population ecology of British shrews in grassland. *Acta Theriologica*, **22**, 279–296.

Pucek, Z., Jedrzejewski, W., Jedrzejewska, B. and Pucek, M. (1993). Rodent population dynamics in a primeval deciduous forest (Bialowieza National Park) in relation to weather, seed crop and predation. *Acta Theriologica*, **38**, 199–232.

Rackham, O. (1975). *Hayley Wood. Its History and Ecology*. Cambridge and Isle of Ely Naturalists' Trust, Cambridge.

Richards, C.G.J. (1985). The population dynamics of *Microtus agrestis* in Wytham, 1949 to 1978. *Acta Zoologica Fennica*, **173**, 35–38.

Shillito, J.F. (1960). The general ecology of the common shrew, *Sorex araneus* L. DPhil Thesis, Oxford University.

Silvertown, J.W. (1980). The evolutionary ecology of mast seeding in trees. *Biological Journal of the Linnean Society*, **14**, 235–250.

Smal, C.M. and Fairley, J.S. (1982). The dynamics and regulation of small rodent populations in the woodland ecosystems of Killarney, Ireland. *Journal of Zoology, London*, **196**, 1–30.

Smyth, M. (1966). Winter breeding in woodland mice *Apodemus sylvaticus* and voles *Clethrionomys glareolus* and *Microtus agrestis*, near Oxford. *Journal of Animal Ecology*, **35**, 471–485.

Southern, H.N. (1970). The natural control of a population of tawny owls (*Strix aluco*). *Journal of Zoology, London*, **162**, 197–285.

Southern, H.N. and Lowe, V.P.W. (1968). The pattern of distribution of prey and predation in tawny owl territories. *Journal of Animal Ecology*, **37**, 75–97.

Southern, H.N. and Lowe, V.P.W. (1982). Predation by tawny owls (*Strix aluco*) on bank voles (*Clethrionomys glareolus*) and wood mice (*Apodemus sylvaticus*). *Journal of Zoology, London*, **198**, 83–102.

Tamarin, R.H. (ed.) (1985). *The Biology of New World Microtus*. Special Publication of the American Society of Mammalogists No. 8, Boston.

Tansley, A.G. (1939). *The British Islands and their Vegetation*. Cambridge University Press, Cambridge.

Tapper, S.C. (1976). Population fluctuations of field voles (*Microtus*): a background to the problems involved in predicting vole plagues. *Mammal Review*, **6**, 93–117.

Taylor, I.R., Dowell, A., Irving, T., Langford, I.K. and Shaw, G. (1988). The distribution and abundance of the barn owl *Tyto alba* in south-west Scotland. *Scottish Birds*, **15**, 40–43.

Tew, T.E., MacDonald, D.W. and Rands, M.R.W. (1992). Herbicide application affects microhabitat use by arable wood mice (*Apodemus sylvaticus*). *Journal of Applied Ecology*, **29**, 532–539.

Toms, M.P. (1990). An investigation into the distribution of small mammal species in different year classes of sweet chestnut coppice. BSc Honours Project, University of Southampton.

Trout, R.C. (1978). A review of studies on wild harvest mice (*Micromys minutus* (Pallas)). *Mammal Review*, **98**, 143–158.

Tubbs, C.R. (1986). *The New Forest*. Collins New Naturalist Series. Collins, London.

Turcek, F.J. (1967). Cycling of some forest tree seeds with special reference to small mammals and then animals in general. In: Petrucewicz, K. (ed.). *Secondary Productivity of Terrestrial Ecosystems*. Panstwowe Wydawnictwo Naukowe, Warsaw, 349–355.

Venables, L.S.V. and Venables, U.M. (1965). Transect trapping of small mammals in conifer plantations, Newborough Warren, Anglesey. *Nature in Wales*, **9**, 171–184.

Watts, C.H.S. (1969). The regulation of wood mouse (*Apodemus sylvaticus*) numbers in Wytham Woods, Berkshire. *Journal of Animal Ecology*, **38**, 285–304.

Williamson, D.R. (1992). *Establishing Farm Woodlands*. Forestry Commission Handbook No. 8. HMSO, London.

Wilson, W.L., Montgomery, W.I. and Elwood, R.W. (1993). Population regulation in the wood mouse *Apodemus sylvaticus* (L.). *Mammal Review*, **23**, 73–92.

Wolk, E. and Wolk, K. (1982). Responses of small mammals to the forest management in the Bialowieza Primeval Forest. *Acta Theriologica*, **27**, 45–59.

Yalden, D.W. (1974). Population density in the common shrew, *Sorex ananeus*. *Journal of Zoology, London*, **173**, 262–264.

Yalden, D.W. (1991). History of the Fauna. In: Corbet, G.B. and Harris, S. (eds). *The Handbook of British Mammals*. Blackwell Scientific Publications, Oxford, 7–18.

13 Do Woodland Mammals Threaten the Development of New Woods?

R. M. A. GILL, J. GURNELL and R. C. TROUT

INTRODUCTION

There is increasing concern for the problems that herbivorous mammals create for woodland management. Although animal damage problems are by no means new to foresters, a number of developments are taking place that are creating new aspects of the problem. Firstly, most of the larger herbivorous mammals (roe (*Capreolus capreolus*), fallow (*Dama dama*) and muntjac deer (*Muntiacus reevesi*), rabbits (*Oryctolagus cuniculus*) and grey squirrels (*Sciurus carolinensis*)) are already increasing in range or numbers in Britain and this may be accelerated by grant-aided initiatives such as the farm and community woodland schemes which extend the area of woodland. Secondly, mounting evidence reveals that deer, if present at too high a density, will decrease the diversity of plants as well as the value of woodland as habitat for many birds, small mammals and invertebrates. This aspect of the impact of deer has taken on greater significance as the multiple uses of woodland have become more important. Finally, techniques for population control and damage prevention currently in use have a number of limitations. For example, fencing and tree guards do not protect against all forms of impact and may be too expensive in situations where conservation is a primary objective. Furthermore, the use of shooting, poisons or fumigation may be impractical for community woodlands near urban areas and public opposition may ultimately limit their wider acceptance.

In this chapter, we review the evidence for the effect that woodland creation will have on populations of herbivorous mammals and for the impact they in turn will have on woodland habitats. We will focus on the species that present the greatest problems in Britain today, which include deer, rabbits and grey squirrels. Although each of these species differs in their effect on woodlands, they are all likely to benefit from woodland creation and present a unified threat, to undermine both the conservation and the economic value of new woodlands. Finally, we discuss the limitations and alternatives to existing control measures.

The Ecology of Woodland Creation. Edited by Richard Ferris-Kaan.

THE INFLUENCE OF NEW WOODLAND ON MAMMAL POPULATIONS

Deer

Five deer species are present in Britain with a substantive range: red (*Cervus elaphus*), roe, fallow, sika (*Cervus nippon*) and muntjac; Corbet and Harris, 1991). All deer can be classified on the basis of diet as either browsers or browser/grazers (Hoffman, 1985), and all make heavy use of woodland habitats.

In the relative absence of shade, young forest stands yield much greater quantities of accessible browse than either older woodland or agricultural habitats (Halls and Alcaniz, 1968). Studies of habitat selection and population dynamics of red and roe deer have revealed that both species benefit from woodland creation, and the same is likely to apply to other species. Where they have access to forests including a range of stand ages, both red and roe deer have been found to make more use of stands less than 15 years old than older stands (Welch *et al*, 1990; Staines and Welch, 1984). Where roe deer have access to recently planted stands, they have been found to achieve higher rates of growth, fecundity and recruitment (Loudon, 1987; Gill, unpublished data) and red deer also achieve higher pregnancy rates and breed at a younger age in woodland in contrast to moorland (Ratcliffe, 1987).

Historically, changes in forest cover, amongst other factors, have made a significant contribution to changes in deer populations. Woodland expansion has been reported to be followed by increases in the range or numbers of roe and red deer in several European countries, including Britain (Argyll and southern England: Prior, 1965; Rowe, 1982), in Scandinavia since the mid-19th century (Ahlèn 1965), and in the Swiss alps and Slovenia since the 1940s (Adamic, 1986; Gill, 1990). Similarly, in the USA, the decline and subsequent recovery of forest area between 1830 and 1980 (Williams, 1989) was associated with a decline and recovery in white-tailed deer (*Odocoileus virginianus*) range and numbers as the influence of forest clearance and human settlement spread from New England towards the west, and later towards the south (Halls, 1984; Gill, 1990).

As woodlands mature and become suboptimal habitat the deer population density is likely to decline, but this may be partly compensated by an altered pattern of habitat use. Woodland edges, adjacent fields or other openings such as rides become the most heavily used sites because they provide the best combination of food and cover (Papageogiou, 1978; Henry, 1981; Thirgood and Staines, 1989; Welch *et al.*, 1990).

Besides forest or woodland, most species of deer are known to be capable of using agricultural habitats (Corbet and Harris, 1991). However, the establishment of significant deer populations in agricultural areas appears to be a relatively recent phenomenon, taking place most rapidly since the 1940s (Gill, 1990). Furthermore, it is mainly roe deer in continental Europe and white-tailed deer in the mid-west USA that have benefited, although fallow and red deer also use farmland in more limited areas. The reasons for this expansion into farmland habitats are not fully understood, but are likely to be related to changes in agricultural crops or practices, and in some areas also to the establishment of woodland. The availability and palatability of crops to deer varies sharply during the course of a year (Johnson, 1984) and therefore the arrangement of crop types and woodland may be important in determining their use by deer.

In spite of their ability to adapt to agricultural landscapes, deer populations in farmland rarely achieve densities as high as those recorded in woodland habitats (Kaluzinski, 1982; Staines and Ratcliffe, 1987; Denis, 1992). The available woodland remains heavily used for both food and cover (Gladfelter, 1984), and range sizes typically increase as deer exploit various crops as they become available (Johnson, 1984). As a result of this dependency on woodland habitats, new woodlands will increase the range and numbers of deer. In farmland or other areas devoid of much tree cover, these woodlands are likely to be small and scattered, creating a large edge to area ratio and providing deer with cover and facilitating the use of nearby fields and crops. The smaller species (muntjac, roe and fallow deer), already present in the lowlands but with a limited distribution, will benefit most directly from this change.

Squirrels

Grey squirrels were first introduced from North America into Britain in 1876, and by 1937 several more introductions and translocations had taken place (Shorten, 1954; Lloyd, 1983). Since the 1930s they have spread rapidly and are now found throughout most of central and southern England and Wales, and over quite large areas of Scotland (Gurnell and Pepper, 1993). They continue to extend their range in northern England and Scotland, and no northern limit to their distribution has so far been detected.

Although grey squirrels are opportunists and eat a wide range of food types, their primary foods are tree seeds (Gurnell, 1991). Consequently tree seed availability has a very important effect on their population dynamics and patterns of distribution (Gurnell, 1983, 1989). Mature broadleaved woods containing large seeded trees such as oak (*Quercus* sp.), beech (*Fagus sylvatica*), sweet chestnut (*Castanea sativa*) and hazel (*Corylus avellana*) are prime habitats for grey squirrels. Here they exhibit high long-term average densities (e.g. 6–8 ha^{-1}). In contrast, conifer woods, or pole stage broadleaved woods are less suitable and hold fewer individuals, e.g. 0.5–1.5 ha^{-1} (Don, 1981; Gurnell and Pepper, 1988; Gurnell, 1989). The size of the seed crop each year has a significant effect on the annual cycle of numbers. When the crop is good, the numbers of adult and juveniles present the following spring and summer are high (Gurnell, 1981, 1983, 1989). If the seed crop fails, the numbers the following spring and summer are low with no juveniles present.

Grey squirrels are well adapted to living in urban woods, parks and gardens and utilise both natural and artificial foods (e.g. food on bird tables and litter bins) found in these habitats. Casual feeding by the public can lead to increases in local densities. Also, when tree seed crops are poor, grey squirrels are more often seen in gardens and it is notable that more damage occurs to soft fruits (strawberries, raspberries, apples) and other garden produce at this time. They can also be a nuisance by raiding dustbins or getting into houses, causing damage to water pipes, cables and roofs.

Grey squirrels take up permanent residence in new woodland if it is of a sufficient size and of an age to provide an all-the-year round food supply. This will usually occur when the main tree species are old enough to produce seed, which ranges from 10 to 50 years depending upon the species. They are also able to commute distances of a kilometre or more on a daily basis (Don, 1981; Kenward, 1985). They will, therefore, make short, foraging visits to some habitats such as young woodland or gardens.

Together, woodlands, parks and gardens provide a patchwork of habitats for grey squirrels of varying quality in terms of food availability (Gurnell, 1981). The juxtaposition of these habitats is important. Patches of prime habitat can be a source of grey squirrels which disperse during the summer or autumn. This particularly occurs in years following a good seed crop when numbers are high. Dispersing animals may then temporarily increase densities in neighbouring young woodlands which are more vulnerable to damage (Don, 1981; Kenward and Parish, 1986; Kenward, et al., 1988; Gurnell, 1989). Grey squirrels move along hedgerows, roadside trees and through small woods and copses (see Fitzgibbon, 1993), but they do not require a continuity of trees and have been known to move between woods several kilometres apart (Shorten, 1954). However, even though discrete habitat patches may be too small to support grey squirrels on their own, networks of small habitat patches interconnected by hedgerows may well support populations and this has significance for certain types of woodland, such as farm woodland (Fitzgibbon, 1993) and Community Forests.

Rabbits

Rabbits originated from the Mediterranean parts of Europe where they occurred in habitats of short arid grassland and scrub. They were probably introduced into Britain by the Normans in the 11th century and kept in formal warren areas. Now, rabbits are found routinely in all parts of the country, with the exception of extremely wet areas and land above 500 m in elevation. After myxomatosis destroyed over 99% of the population in 1954–6 (Thompson and Wordon, 1956), the recovery of the rabbit population was slow and initially very patchy. Rabbits are again widespread (Trout et al., 1986) and damage in agricultural areas is estimated at over £100 million and rising (Rees et al., 1985). The effect of myxomatosis has weakened since the 1950s as rabbits became resistant to the virus, but it is still known to be a major cause of mortality (Trout et al., 1992).

Rabbits have two basic requirements; a place to dig and maintain a burrow system in which the young are born and reared, and feeding areas preferably containing a predominance of short and nutritious vegetation. In much of Britain, burrows are confined largely to woodland and field edges, where the physical disturbance by intensive agriculture is minimal. A regression analysis of land-use, soil and meteorological factors from 450 km^2 squares in England revealed that rabbits were more abundant in those areas containing a high proportion of woodland ($t = 4.99$, 248 df, $p<0.001$; Trout et al., in prep.). Unfortunately, few squares contained more than 25% woodland so it is unclear whether this relationship holds in areas that are predominantly forest or woodland. Further, a principal component analysis of the habitat variables suggested that the distribution of the woodland would also be important; those sites with many small patches of woodland and a long length of woodland edge habitat tended to have higher rabbit abundance scores ($r = 0.271$, $p<0.001$). In large unbroken blocks of woodland, warrens are found usually within a short distance from the perimeter, adjacent to the food resource. Burrows may be found deeper into the woodland, but these are usually near open areas with an associated food supply.

The rabbit is a generalist herbivore, requiring about 100–120 g dry vegetation per day (Monk, 1987), though this figure is probably very variable in practice. Its diet is completely dependent upon the nature of the available food supply within a few hundred metres of their warren. Rabbits prefer highly nutritious graminaceous material for the bulk of food but take a wide range of species, and can survive on extremely poor food in winter (e.g. purple moor grass (Molinia caerulea), fescues (Festuca sp.) and heather

(*Calluna vulgaris*)), competing easily with sheep. Before the outbreak of myxomatosis, classical studies revealed that rabbit grazing at densities of 7–25 ha^{-1} prevented scrub regeneration on chalk downland and Breckland, demonstrating that small tree material is readily consumed (Tansley, 1949; Thomas, 1960; Smith, 1980). Rabbits are again now widespread (Trout *et al.*, 1986) and grazing of woody plants common (Lloyd, 1970; Sumption and Flowerdew, 1985). Most young plantations on farmland now have some protection against rabbits and there is also an increasing trend by the UK Forest Enterprise to protect new plantings against them (H. W. Pepper, personal communication).

Rabbit populations are not likely to recover to the pre-myxomatosis levels everywhere, because of the removal of so many hedgerows since the 1950s, but appear close to these levels locally. New woodlands in the agricultural landscape provide an extra area of habitat for invasion, population growth and then impact on local vegetation. The effect on rabbits of creating new woodland will be profound. Recent initiatives to plant woodlands appear to be creating habitat for new burrow systems and unless appropriate controls or protective measures can be taken, rabbit numbers will increase causing damage both to young trees and agricultural crops.

DAMAGE TO TREES

Deer

All species of deer can cause damage to young trees by browsing leaves or shoots and by fraying with antlers. In addition, the larger species (red, sika and fallow deer) damage saplings and pole-sized trees by stripping bark. Typically, deer damage shows a marked seasonal pattern, with most bark stripping and browsing on evergreens taking place in winter and browsing on deciduous species in summer (Gill, 1992a). Fraying is normally most intense before the rut or, in the case of roe, also during territory establishment in the spring (Sempéré *et al.*, 1980). The severity of each of these types of deer damage can vary enormously from area to area. Broadly, there are three main factors that affect the amount of damage: the numbers of deer, their patterns of habitat selection and their choice of food plants. In addition, tree species differ in their response to damage; this aspect is dealt with in the next section.

The severity of damage is widely recognised to increase with population density (Holloway, 1967; Maizeret and Ballon, 1990; Welch *et al.*, 1991). In most cases, however, the relationship between damage and density is poorly defined, making it difficult to make use of the data to assess the likely benefit of control. A clear example has been provided by Tilghman (1989), who kept white-tailed deer in enclosures at densities ranging from 4–31 km^{-2} At a density of about 15 km^{-2} or above, browsing began to have a significant impact, reducing the height and diversity of seedlings. The existence of a threshold below which little or no damage occurred is encouraging in the sense that it suggests that damage can be virtually eliminated if control successfully maintains numbers below this level. Although few other studies have made such a direct comparison of damage at different densities the evidence suggests that this level is very variable. Holloway (1967) and Stehle (1986), for example found little damage by roe deer at densities of 6–10 km^{-2} although in central France the survival of *Quercus* seedlings has been found to be unaffected by densities as high as 25 km^{-2} (Ballon *et al.*, 1992; Bideau *et al.*,

1992). Where red deer occur in Scotland, damage may become significant at densities of 4 km^{-2} or even less (Holloway, 1967). In some cases, the lack of concurrence between damage and density is due to the effects of habitat selection. Sites offering the best combination of food and cover often attract the greatest damage, for example at the edges of clearings and rides (Thirgood and Staines, 1989). This effect is greater where roe deer are present than where red or fallow occur, because roe deer have smaller home ranges and are less likely to wander far from cover (Thirgood and Staines, 1989; Kay, 1993). Heavy browsing of ride and clearing edges creates severe problems for forest ride management, where the objective is to encourage good broadleaved shrub cover along the edges (Anderson and Buckley, 1991). The influence of hiding cover may be countered where disturbance by people or traffic is high (Repo and Loyttyniemi, 1985), although culling may need to be maintained even in these areas to prevent deer from becoming too tame.

Deer are very selective feeders and this is clearly reflected in the pattern of browsing or bark stripping, with trees of particular sizes, species, age classes and even varieties receiving more damage than others. The most heavily browsed broadleaved species include willows (*Salix* spp.), oak (*Quercus* spp.), ash (*Fraxinus excelsior*) and rowan (*Sorbus aucuparia*), whereas silver fir (*Abies alba*), Douglas fir (*Pseudotsuga menzesii*) and larch (*Larix* spp.) are the most susceptible conifers. The least palatable species include alder (*Alnus glutinosa*), Sitka spruce (*Picea sitchensis*) and Corsican pine (*Pinus nigra* var. *Maritima*) (Mitchell *et al.*, 1977; Gill, 1992a; Kay, 1993; Putman, 1994). In spite of generally well-recognised differences in palatability between species there is some regional variation, which may be due to differences in the vegetation, soil fertility or tree varieties. Norway spruce (*Picea abies*) and Scots pine (*Pinus sylvestris*), for example, often attract little damage from roe deer in southern England although this is not usually the case in the uplands (Rowe, 1982).

Trees species do not exhibit the same relative vulnerability to bark stripping. In general, smooth or thin-barked species such as *Salix* spp., ash and rowan are more readily stripped than species with thick, rough or flaky bark, such as birch, oak or Sitka spruce (Gill, 1992a). In most species, bark increases in thickness and roughness with age resulting in a decline in vulnerability. The rate at which bark thickens varies between species; in lodgepole pine (*Pinus contorta*) and Scots pine vulnerability ceases at about 16 years of age but it extends up to 50 years in Norway and Sitka spruce (McIntyre, 1975; Welch *et al.*, 1987).

Deer damage can be affected by the availability of alternative browse plants. There are several reports of damage being reduced where more palatable vegetation is available (Huss and Olberg-Kolfass, 1982; Furrh and Ezell, 1982; Eiberle and Bucher, 1989). However, sites offering good feeding may attract more use, leading to more damage (Holloway, 1967; McIntyre, 1975). In view of the fact that the relationship between damage and other vegetation can be difficult to predict, attempts to reduce damage by providing food are not generally recommended, and may be counter-productive by stimulating recruitment into the deer population. Limited feeding may, however, be useful for drawing deer away from vulnerable areas or for creating viewing opportunities. This is most likely to be successful with red or fallow deer, which range over much larger areas and are more likely to be drawn further than roe deer or muntjac (Gill, 1992a).

Where deer numbers are high, browsing on any unprotected young tree should be expected, even the least palatable species. Deer have often been reported to switch to less palatable species after the most palatable ones have been depleted (Beals *et al.*, 1960; Stewart and Burrows, 1989). Furthermore, in the absence of predation or culling, the density-dependent effects on recruitment or adult survival will not prevent densities rising above the level at which damage or some other unwanted impact becomes significant. In areas with little or no deer control, red deer may achieve densities as high as 40 km^{-2} and roe deer may exceed 75 km^{-2} (Ratcliffe, 1984; Gill, unpublished data), well above the range of 4–25 km^{-2} reported at the beginning of this section, for areas where damage was considered acceptable.

Grey squirrels

Squirrels damage trees by biting or pulling off the outer bark, either in the crown or main stem, and exposing the sap which may then be eaten (Shorten, 1957; Mackinnon, 1976) (Table 13.1). Many broadleaved species can be damaged by grey squirrels and there is now some concern about damage occurring to conifers. However, the most vulnerable species are beech and sycamore with poplar (*Populus* spp.), ash, silver birch and oak damaged less often (Rowe and Gill, 1985; Gill, 1992b). Normally, damage occurs when the trees are between 10 and 40 years old, but damage to small saplings can also occur (Gill, 1992b). Trees tend to be most susceptible when they are growing most vigorously (Kenward and Parish, 1986), and provided the stand is within the vulnerable age band, this often results in selection for the largest trees in the stand (Mackinnon, 1976). The main damage period in Britain is between April and July when the sap is rising and the bark is easy to remove (Hampshire, 1985; Gill, 1992b).

Table 13.1. Grey squirrel bark stripping damage to vulnerable tree species in state and private forests in the UK in 1983. From J.J. Rowe, 1984, *Quarterly Journal of Forestry*, **78**, 231–236, and reproduced by permission from the Royal Forestry Society of England, Wales and Northern Ireland. Damage assessment defined as the presence of exposed wood due to bark removal by squirrels

Species	Number of areas		Damage (% trees attacked)					
	Assessed	Damaged	0	1–20	21–40	41–60	61–80	81–100
Beech	143	87	13	39	17	17	13	<1
Sycamore	118	86	14	39	20	13	9	5
Oak	139	41	59	31	6	1	2	<1
Poplar	45	29	71	13	4	9	0	2

Damage by grey squirrels is usually most severe when densities are highest in woodlands containing vulnerable trees, typically pole stage stands (Kenward and Parish, 1986). Densities can be strongly influenced by a good mast crop the previous autumn (Gurnell, 1989) and therefore the proximity of mature trees to pole stage woodland can be an important determinant of the amount of damage (Gurnell and Pepper, 1988; Gurnell, 1989; Kenward *et al.*, 1988).

Although grey squirrel damage occurs at the time of the year when their natural foods (tree seeds, leaf buds and flowers, and fungi) are scarce, stripping does not appear to be caused by a food shortage because it can still occur close to food sources, such as poison hoppers (H.W. Pepper, personal communication). In view of this, damage is thought to be influenced by agonistic behaviour between dominant and subordinate individuals (Taylor, 1969; Gurnell, 1987) which may in turn be exacerbated by competition for food or other resources. However bark stripping and associated behaviour has proved to be difficult to observe in the wild and a complete explanation of the motives for it remain elusive (Gill, 1992b).

Rabbits

Rabbits have long been known to be capable of causing serious damage to woodlands. Before the outbreak of myxomatosis, reports reveal that repeated efforts at planting, in conjunction with both intensive control and fencing, were sometimes necessary to successfully establish woodland (Smith, 1954; Sheail, 1971). At that time, the total cost to private forests was estimated at £1.5 million (Thompson and King, 1944). Now that numbers have largely recovered, complete losses are being reported again on unprotected new plantings and rabbit management has become a significant component of woodland management; 30–50% of the establishment costs of a Farm Woodland Premium Scheme relate to protection from rabbits.

Rabbits damage trees in several ways: by biting through the stem or browsing off side shoots of young trees, by removing the bark or by digging and eating the roots. Small stems are cut at a characteristic 45° angle, clearly different to the 'tear and tail' produced by deer. Cut stems are frequently not consumed, particularly in farm woodland plantings; bark removed is, however, completely eaten, unlike the large shreddings left by squirrels. Damage is usually limited to small trees, but they sometimes also strip bark from mature trees of smooth-barked species like beech and ash, particularly during prolonged snow conditions. Rabbits can reach to about 60 cm from the ground (or from the top of snow cover) so physical damage is usually restricted to below this level.

The reasons for the different forms of damage by rabbits, particularly the habit of clipping and abandoning young trees, are not entirely understood. Bark stripping in the winter may be done to obtain carbohydrates stored in the stem, but severe damage can occur in summer as well. The provision of an additional or alternative food source appears to have little effect on damage, particularly if the density of rabbits is too high. In experimental trials within paddocks, rabbits at a density of 10 ha^{-1} or more have attacked and destroyed unacceptable proportions of ash and beech seedlings, in spite of having access to supplementary food (Figure 13.1). On former agricultural land the ground vegetation is high in nutrients and will attract rabbits, thus putting unprotected trees at greater risk.

Tree species have been known for some time to differ in susceptibility to rabbit damage (Michie, 1869). More recently, marked differences between provenances of both birches and willows have also been found (Trout and MacVicker, 1993). The species that appear to be most palatable include larch, ash, willows and beech while those that are not damaged so often, or so severely, include elder (*Sambucus nigra*), buckthorn (*Rhamnus catharticus*), balsam poplar (*Populus balsamifera*) and sycamore. However, most species planted are likely to be destroyed if the population pressure is very high.

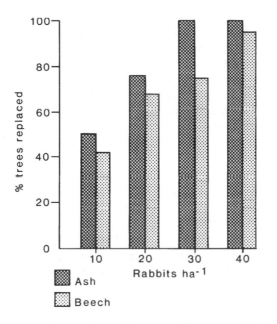

Figure 13.1. The impact of increasing rabbit (*Oryctolagus cuniculus*) density on young unprotected trees. Sixty trees of each species were planted (unprotected) into 0.4 ha paddocks, containing rabbits maintained at densities of 10, 20, 30 and 40 individuals ha⁻¹. The results represent the average proportion (from six replicated trials) of trees damaged severely enough to need replacing after one year

TREE SURVIVAL, GROWTH AND TIMBER QUALITY

Browsing

Browsing by deer and other mammals can delay growth, increase the likelihood of death or reduce the quality of timber by resulting in stem deformations. In general, both the amount of growth loss and likelihood of death are dependent on the severity or frequency of browsing damage (Holloway, 1967; Bergstrom and Danell, 1987; Eiberle and Nigg, 1987). The age or size of the tree is likely to determine whether browsing kills the tree or simply reduces growth. Younger or smaller trees are more likely to die than older or larger ones (Krefting and Stoeckeler, 1953), although the age at which mortality declines to near zero is known to vary between at least 1–3 years, depending on the tree species and environment (Holloway, 1967; Lewis, 1980). Once established, trees can withstand frequent or repeated damage with little effect on survival.

The difference between tree species in their susceptibility to browsing is substantial, and may be as important as their palatability in determining the influence of browsing on tree species composition (Gill, 1992c). The ability of a tree to recover depends on the distribution of dormant buds that can be activated after shoot loss, as well as the time of year that damage occurs. The species that are found to recover well from browsing include birches, willows, rowan, oak and aspen (*Populus tremula*); in contrast, Scots pine and silver fir are usually found to be more vulnerable (Eiberle, 1975; Miller *et al.*, 1982).

After severe browsing, trees typically show some compensation, growing relatively faster than before the damage occurred. Although few studies have been done of the long-term effects of browsing on growth, most suggest that compensation is only partial, and therefore some permanent growth loss remains (Gill, 1992c). A growth delay is likely to be most serious in plantings on former agricultural land where competition from weeds is intense and prolonged use of herbicide application is required. In addition to growth loss, browsed trees can develop multiple stems or a severe sweep in the stem as a result of the loss of the leading shoot (Eiberle, 1975; Welch et al., 1992). Although there can be an improvement in stem form as the leaders recover dominance, leader damage results in a greater proportion of thinner, curved stems and therefore a loss of timber volume. Recent evidence suggests that this is likely to be a far more serious consequence of damage than growth loss (D. Welch, personal communication).

Environmental factors are known to be critical in determining growth and survival after damage. The combination of shade and damage has been found to reduce seedling survival more than either shade or damage alone (Crouch, 1966), indicating that browsing will eliminate trees under partial shade much more quickly than in the open. Similarly, the combination of browsing and competition from other plants is likely to be much more detrimental than damage to trees growing free of competition.

Bark stripping

In addition to browsing, bark stripping could also have serious consequences for trees in new woodlands. Bark stripping will kill trees if the entire circumference is removed from the main stem below the live crown. Although severe attacks by grey squirrels and rabbits sometimes do this to young trees, the majority of bark wounds result in only partial girdling. These wounds none the less result in deformations and sometimes form an entry point for fungal attack, leading to stain, decay and weakening of the main stem (Pawsey and Gladman, 1965; Gregory, 1986). The amount of timber lost by decay depends largely on the size of the wound and the species of the tree (Löffler, 1975; Gregory, 1986). In species normally damaged by grey squirrels in the upper main stem (oak, poplars and conifers), subsequent gales sometimes break the crown at the site of the wound, either killing the tree or leaving it severely misshapen. In view of the fact that trees are usually vulnerable to bark stripping for many years, fencing against deer and rabbits is not usually cost effective and damaged trees are too mature for replacement. As a result, bark stripping can be a difficult and sometimes costly problem in forest management.

THE IMPACT ON NATURAL REGENERATION

Deer

Deer are widely reported to reduce the size and number of regenerating seedlings (Gill, 1992c). In general, the effect increases with increasing animal density, and can result in the virtual elimination of all seedlings when browsing pressure is too high. Putman et al. (1989) found that an area protected from fallow deer had a density of 6440 saplings ha^{-1} after 14 years, in contrast to only 20 ha^{-1} in an adjacent heavily browsed area. The loss of trees and shrubs is likely to be more marked under partial shade, which limits the capacity of trees to recover from browsing.

Deer populations have been increasing in much of Europe and North America during the 20th century, and as a result, problems have developed in both continents in forests managed by natural regeneration. The most commonly reported problem is a change in the composition of seedlings, with the species most heavily browsed, or least able to recover from damage disappearing, often at the expense of the most resistant species. In central Europe, for example, silver fir is declining in abundance and in the Great Lakes region Eastern hemlock (*Tsuga canadensis*), white cedar (*Thuya occidentalis*) and Canadian yew (*Taxus canadensis*) are declining, often at the expense of sugar maple (*Acer saccharum*) (Beals *et al.*, 1960; Kammerlander, 1978; Zeltner, 1979; Frelich and Lorimer, 1985; Alverson *et al.*, 1988). Where several species form the canopy, browsing usually results in a decline in tree species diversity (Marquis and Brenneman, 1981; Tilghman, 1989; Jones *et al.*, 1993).

In spite of the suppression of regeneration that occurs in the presence of deer browsing, sufficient regeneration to establish a tree crop can still be possible where seedling density is initially high and deer numbers are not excessive (Barandun, 1983; Brandner *et al.*, 1990; Heikkilä, 1992; Ballon *et al.*, 1992; Lyly and Saksa, 1992). Seedling recruitment as a function of both deer and tree density has been more widely investigated for moose (*Alces alces*) browsing on conifers than other deer species. Where seedlings are dense, the palatability and average rate of browsing per tree declines (Thompson, 1988; Heikkilä and Mikkonen, 1992), resulting in recruitment of a sufficient density of seedlings. In these cases, the initial density of seedlings needed to be in the range of 4000–12 000 ha^{-1}, but many more than this may be required in other circumstances. The success of regeneration in the presence of even moderate deer densities is likely to be possible only where the other requirements for regeneration, including light intensity, freedom from competition from other plants and seed supply, are ideal.

Although not of immediate concern in the creation of new woodland, the long-term consequences of deer browsing need to be considered, because they emphasise that deer control may be necessary beyond the establishment phase. The removal of young shrubs and trees will result in a simplification of the vegetation structure, resulting in a loss of habitat for wildlife. Ultimately, the failure of recruitment will lead to the depletion of sensitive species altogether and restoration of structural and species diversity may take many decades after deer numbers have been reduced (Anderson and Katz, 1993).

Squirrels

There has been a considerable amount of debate as to whether grey squirrels, in common with other rodents and some birds, enhance or reduce natural regeneration. On the positive side, grey squirrels disperse and cache seeds of many species of tree (Smith and Follmer, 1972; Whelan *et al.*, 1991; Worrell and Nixon, 1991). Some seed caches may not be recovered and result in seed germination (Stapanian and Smith, 1986; Thompson and Thompson, 1980). Furthermore, seed burial can provide ideal germination conditions (Stapanian and Smith, 1986; Tomback, 1982; Jensen, 1985). On the negative side, however, is the considerable amount of evidence which suggests that consumption of seeds by grey squirrels and other seed 'predators' is an important factor in reducing or preventing natural regeneration in many areas (Ashby, 1959; Tanton, 1965; Shaw, 1968; Nilsson, 1985).

The success of regeneration is therefore likely to depend on the irregularity of good seed crops (Silvertown, 1980; Fenner, 1991; Gurnell, 1993). In poor seed years it is likely that seed predators will remove most if not all of the seeds, but in mast years predator satiation will result in some seeds, whether cached or not, escaping predation and germinating (Shaw, 1968; Jensen, 1985; Gurnell, 1993).

Unfortunately these studies give little indication of how much grey squirrels affect regeneration over and above the effects of other seed predators. Recent evidence, however, suggests that they have a largely negative influence and in some cases may be responsible for the failure of natural regeneration. A study by Pigott et al. (1991), for example, has shown that grey squirrels excise the radicle of the acorns they bury which prevents germination even if they are not reclaimed. Further, they have also been reported to damage younger saplings (Ormrod, 1991), and the combined effects of this with seed and seedling predation can eliminate natural regeneration of beech and sycamore (Pigott et al., 1991). In many small woods and urban parks where no squirrel control is practised, saplings of these species are disappearing before the trees reach 30 years of age, resulting in an age structure containing only very young and mature or over-mature trees (H.W. Pepper, personal communication).

Rabbits

There are many reports of rabbits having a negative impact on woodland regeneration. The majority, however, date from the period immediately preceding the onset of myxomatosis, when they were clearly limiting regeneration in many areas (Macdonald and Lockhardt, 1953; Kennedy, 1956). Following myxomatosis, bursts of regeneration of many tree and shrub species were reported, particularly in areas of dune and downland vegetation where rabbit numbers had been especially high (Thomas, 1960; Fuller and Boorman, 1977; Sumption and Flowerdew, 1985). Few authors report a selective effect on tree regeneration, possibly because rabbits tend to have an intense and rather local impact, but Miles (1972) and Haber and Matuszewski (1966) noted that oaks appeared to be more susceptible than many other tree species.

IMPACT ON WOODLAND GROUND VEGETATION

Mammalian herbivores are well known to be capable of having a dramatic effect on the structure and composition of vegetation (Crawley, 1989). Almost all herbivores are required to feed selectively to maximise digestive efficiency, and because competition has such a powerful influence in shaping plant communities (Crawley 1990), even a modest level of selective feeding can result in a change in species composition. Apart from their influence on regeneration, grey squirrels are not thought likely to influence woodland vegetation structure and have therefore been excluded from the discussion in this and the following section.

Deer

The species which almost invariably show decline in response to deer browsing include bramble (*Rubus fruticosus*), honeysuckle (*Lonicera periclymenum*), ivy (*Hedera helix*), wild rose (*Rosa canina*) and wide range of shrubs and young broadleaved trees, including holly (*Ilex aquifolium*), gorse (*Ulex* sp.), heather (*Calluna vulgaris*), willows, oaks and ash. Those that increase include unpalatable species or those most tolerant of damage, such as bracken (*Pteridium aquilinum*), grasses and mosses as well as herbaceous plants that may be toxic, such as foxglove (*Digitalis purpurea*), ragwort (*Senecio jacobea*), horsetails (*Equisetum arvense*) and ground ivy (*Glechoma hederacea*) (Picard, 1976; Kraus, 1987; Cibien *et al.*, 1988; Putman *et al.*, 1989, author's data).

Strikingly similar changes have also been reported in North America in response to browsing by both mule deer (*Odocoileus hemionus*) and white-tailed deer. The cover of grasses and unpalatable ferns increase at the expense of *Rubus*, broadleaved trees and shrubs (Horsley and Marquiss, 1984; Tilghman, 1989; Trumbull *et al.*, 1989). These results suggest that deer are likely to have a comparable effect on woodland regardless of the species of deer or woodland type.

In addition to these changes there is widespread concern that deer browsing reduces the numbers and diversity of herbs and flowering plants. Herbaceous plants form a significant although varied component of the diet of deer (Holisovà *et al.*, 1992) and competition from grasses may help to reduce or even eliminate those species that remain. However, because deer focus much of their browsing on young trees, shrubs and climbers they effectively retard woodland succession, helping to maintain openings (Peterken and Jones, 1989) and reduce shading, thus prolonging the existence of a ground layer. In permanent openings and rides the shading effect is irrelevant and deer feeding is likely to result in a reduction in species richness (Jones *et al.*, 1993). Furthermore, in some areas, nationally rare plants, such as the oxlip (*Primula elatior*) are known to be depleted by deer browsing (Rackham, 1975). More evidence is none the less needed on the loss of diversity and which plant species are affected, especially where several species of herbivore are present.

Rabbits

The great majority of studies of the impact of rabbits on vegetation have focused on grassland or heathland, where pressure has historically been greatest. However, the results are likely to be generally applicable to woodland openings.

Many of the vegetation changes caused by rabbits are comparable with those caused by deer, with the important distinction that they focus more of their feeding on grasses than deer. Changes in the species composition of grasses are very apparent, with the most palatable species decreasing at the expense of the least palatable. Under heavy grazing pressure, flowering plants including orchids also decrease, and toxic species, such as *Senecio jacobaea* and *Glechoma hederacea*, may increase (Thomas, 1963). The intensity of rabbit grazing decreases rapidly with distance from the burrow systems, and this effect helps to maintain openings and increase spatial diversity. They also exhibit latrine behaviour, providing sites suitable for nutrient demanding species such as stinging nettles (*Urtica dioica*), and elder (*Sambucus nigra*).

THE IMPACT ON WOODLAND FAUNA

In view of the practical difficulties involved, there are very few studies specifically comparing animal populations between areas open to grazing or browsing with enclosures large enough to provide habitat for animals. Some inferences can be drawn from the vegetation changes that deer or rabbits create (Plate 13.1), coupled with a know-ledge of the habitat requirements of the animals themselves, although this should not be regarded as a substitute for direct evidence.

The reduction in understorey vegetation caused by deer browsing has a marked effect on invertebrate, bird and small mammal populations. Further changes, mainly to insect and bird populations, will be caused if browsing results in a change in tree species composition. A less direct, but important threat is also posed to species such as the dormouse (*Muscardinus avenallarius*) and some Lepidoptera that depend on coppice woodland, from which it can be difficult to generate sufficient income to justify fencing against deer or rabbits.

Invertebrates

Putman *et al.* (1989) compared the abundance of invertebrates caught in pitfall traps between areas exposed to browsing by *D. dama* with an enclosure which had been protected for 22 years. The area free of browsing developed higher populations of most families of Coleoptera, Diptera and Phalangida (harvestmen), although staphylinids and some species of carabid beetle were more abundant in the browsed area. Two other main groups which were also more abundant in the browsed area included ants and true spiders (Araneae).

So far, no published study appears to have investigated the effect of deer browsing on Lepidoptera. Woodland butterfly populations have declined alarmingly in recent years (Warren, 1992; Warren and Thomas, 1992) largely due to the loss of coppice and hedgerows. However, of the 17 species that Dennis (1992) lists as occurring primarily in woodland habitats, the majority should be considered as potentially at risk from the effects of deer because their larvae feed on trees, shrubs or herbaceous plants that usually decrease under the influence of deer browsing. Only three species, the speckled wood (*Pararge aegeria*), ringlet (*Aphantopus hyperantus*) and heath fritillary (*Mellicta athalia*) which feed on grasses or other unpalatable plants, are likely to benefit from deer.

Apart from changes in vegetation structure, changes in tree species composition are also likely to have an effect over the long term on invertebrate fauna. Tree species differ considerably in the number of phytophagous insects known to be associated with each of them (Winter, 1983; Kennedy and Southwood, 1984). Some species, such as oak and willows, are known to support a wide range of invertebrates but are among the most vulnerable to browsing by both deer and rabbits.

Birds

Browsing may also have a marked influence on the bird community. A reduction in tree species richness or a loss of broadleaved species like oak is likely to be detrimental to woodland bird populations (Fuller, 1982). Furthermore, in view of the fact that deer reduce the height of the shrub understorey, the suitability of the woodland as habitat for many bird species is reduced still further. An area of mature oak woodland in the Forest of Dean, protected from sheep grazing for 45 years, was found to contain higher densities of overwintering passerines than either grazed oak or conifer woodland in the same forest, and was the only site where a nationally rare species, the hawfinch (*Coccothraustes coccothraustes*) was recorded (Hill *et al.*, 1991).

Plate 13.1. The impact of browsing by rabbits (*Oryctolagus cuniculus*) and roe deer (*Capreolus capreolus*) on ride edge vegetation, Micheldever Forest, Hampshire, England. Five years after erecting the fenced plots, tree and shrub regeneration inside the exclosures (on the left of the photograph) was prolific, whereas it was sparse outside, with the vegetation dominated by grasses (R. Ferris-Kaan, The Forestry Authority)

Small mammals

The reduction in vegetation cover brought about by cutting or grazing can be detrimental to small mammal populations (Flowerdew and Trout, Chapter 12). Putman *et al.* (1989) found that the area protected from grazing by fallow deer supported good populations of woodmice (*Apodemus sylvaticus*) and bank voles (*Clethrionomys glareolus*) as well as small numbers of shrews (*Sorex* sp). and yellow-necked mice (*Apodemus flavicollis*). In contrast, a heavily grazed area supported only a reduced number of woodmice. On open ground in woodland, deer browsing, by encouraging grasses, may result in higher densities of field voles (*Microtus agrestis*), but this does not yet appear to have been demonstrated.

DAMAGE LIMITATION

Clearly, deer, grey squirrels and rabbits can have a deleterious influence on woodland, whether looked at from the viewpoint of timber production or wildlife conservation. In the absence of control, they are all likely to achieve densities that will result in an undesirable impact, and natural regulation cannot be relied on to limit numbers. Several options, based on either protection or control, are already available for preventing or limiting the damage that can be done by deer and other mammals. Unfortunately many of these methods have certain limitations, including being expensive, and new approaches need to be sought.

Protection

Well-developed techniques exist for protecting trees against deer and rabbits with fencing or tree guards or repellents (Pepper, 1977, 1978, 1992; Potter, 1992). If well maintained these methods are effective but have the disadvantage of being costly and unsightly, and fencing has the further disadvantage of restricting recreational use of woodland. They are also difficult to use if small felling coupes for natural regeneration are desired. Furthermore, tree guards and repellents only protect trees, not the surrounding vegetation. Although fencing can protect both, the effect of complete exclusion may be as undesirable as high deer or rabbit numbers, for example by inhibiting regeneration due to excessive competition from other vegetation (Pigott, 1983).

Investigations are being carried out to explore the feasibility of developing tree varieties resistant to herbivore damage. Natural variation in palatability has been found among varieties of several tree species to damage by deer, rabbits and other mammals (Gill, 1992a,b; Trout and MacVicker, 1993), and therefore the potential exists to breed for resistance (Rousi, 1990). This approach is however likely to involve long-term research and, in common with individual tree protection, will not protect other vegetation.

Control

From the point of view of overall levels of impact, the most satisfactory approach is to reduce population density by control or, if feasible, by encouraging predators. Deer control is practised in most forests in Britain and is sometimes used in conjunction with other forms of protection (Ratcliffe, 1987; Ratcliffe and Mayle, 1992). However there is still a lack of awareness among the general public as well as landowners of many of the detrimental effects deer have on woodlands. The problem is complicated where deer move across landowners' boundaries, with differing or even opposing management objectives. Initiatives to encourage neighbourhood deer management schemes are badly needed, preferably in conjunction with government support.

Within many private and public British forests grey squirrel numbers are controlled to reduce tree damage. Few are now shot or trapped and these are not efficient or economic methods of control. Control of grey squirrels is almost now entirely achieved by poisoning them with bait containing 0.02% warfarin dispensed in hoppers (Gurnell and Pepper, 1988). It is illegal to poison red squirrels (*S. vulgaris*) or to poison grey squirrels where red squirrels are at risk. In recent years, considerable improvements to hopper design have been made and risks to non-target species have largely been removed (Pepper, 1989a,b, 1990). Furthermore, tests are currently under way on another design of hopper which hopefully will eliminate the risks to red squirrels. Such hoppers will not only be used to reduce tree damage by controlling grey squirrel numbers but they will also be incorporated into strategies to conserve red squirrels (Gurnell and Pepper, 1993).

For rabbits, fumigation of burrows is usually the most effective direct population control technique. Shooting and ferreting, though often used, are less efficient, but trapping can sometimes be effective, particularly in limited areas (Rees *et al.*, 1985). Unfortunately, the principal legal control methods that have been developed for mammals, including shooting, fumigation and poisoning for deer, rabbits and grey squirrels, respectively, may become increasingly objectionable to the public. Furthermore they may simply prove to be impractical in areas of high public access such as Community Forests and the urban fringe.

A possible approach to controlling mammal pests would be to encourage predators, or at least cease any form of predator control. There is increasing evidence that predators, both of small and large mammals, can limit prey population densities. Further, in circumstances where predator dispersal is not restricted and alternative prey species are also available then predators may regulate prey populations at low levels, maintaining them within limits which are unlikely to result in severe damage (Newsome, 1990; Skogland, 1991). Trout and Tittensor (1989) showed that sites with a predator removal policy had twice the abundance index of rabbits than sites with no removal. They argued that although predators were unlikely to reduce high populations, the effect would be to extend the period of low population once some other agent had lowered rabbits below a threshold, thus saving farmers considerable cost in annual management.

Unfortunately, there are limitations with encouraging predators, particularly since it can be difficult to ensure that the predator focuses enough on the intended prey and avoids domestic animals or rare species. Also grey squirrels, being arboreal, can escape many predators and their populations may not be affected much by predation (Gurnell, 1987). Livestock rearing imposes a severe constraint to the conservation of large predators, although in some parts of continental Europe and North America, measures are being taken to encourage or re-introduce wolves (*Canis lupus*), cougars (*Felis concolor*) and lynx (*Felis lynx*), particularly where conflict with livestock will be minimal (Piechocki, 1994).

Predators therefore offer some scope for limiting herbivore problems and deserve more consideration than they attract at present. However, it may prove to be difficult to reconcile predator conservation measures with other land use interests.

CONCLUSIONS

Woodland mammals are clearly a threat to forest and woodland vegetation. If left unchecked, browsing and bark stripping by deer, rabbits and grey squirrels can cause a growth delay and serious loss of revenue from stem deformities, decay or death of small trees. Deer and rabbits may also cause a reduction in taxonomic and structural diversity of trees and shrubs, resulting in the loss of habitat for a wide range of invertebrates, birds and mammals. Although protection in the form of guards and fencing is readily available these techniques do not protect against all forms of impact, nor against grey squirrels, and may more than double the cost of tree establishment. Moreover, the widespread use of poison or fumigation to control grey squirrels or rabbits, respectively, may not be generally acceptable. As a result, excessive browsing and bark stripping by mammals will reduce the economic value of forests and woodlands, restricting future investment in forestry and woodland initiatives, as well as limiting the wildlife value of forests and coppice. The combination of increasing numbers and distribution of deer, grey squirrels and rabbits, coupled with concern for public acceptance for established methods of control, therefore poses a severe threat to the development of new woodlands.

ACKNOWLEDGEMENTS

The authors would like to thank P.R. Ratcliffe and H.W. Pepper for helpful comments on the manuscript, and the Land Use, Conservation and Countryside Group, MAFF, for financial support.

REFERENCES

Adamic, M. (1986). The land use changes in Slovenia and their influence on range and density of some (game) wildlife species. *18th IUFRO Congress*. Division 1, vol. II, Ljubljana, Yugoslavia, September 7–21, 1986, 588–600.

Ahlen, I. (1965). Studies on the red deer (*Cervus elphus*) in Scandinavia. III Ecological investigations. *Viltrevy*, **3**, 177–376.

Alverson, W.S., Waller, D.M. and Solheim, S.L. (1988). Forests too deer: edge effects in Northern Wisconsin. *Conservation Biology*, **2**, 348–358.

Anderson, M.A. and Buckley, G.P. (1991). Managing edge vegetation. In: Ferris-Kaan, R. (ed.). *Edge Management in Woodlands Alice Holt, Surrey 1989*. Forestry Commission HMSO, 5–7.

Anderson, R.C. and Katz, A.J. (1993). Recovery of browse-sensitive tree species following release from white-tailed deer (*Odocoileus virginianus* Zimmerman) browsing pressure. *Biological Conservation*, **63**, 203–208.

Ashby, K.R. (1959). Prevention of regeneration of woodland by field mice (*Apodemus sylvaticus*) and voles (*Clethrionomys glareolus* Schreber and *Microtus agrestis* L.). *Quarterly Journal of Forestry*, **53**, 228–236.

Ballon, P., Guibert, B., Hamard, J.P. and Boscardin, Y. (1992) Evolution of roe deer browsing pressure in the forest of Dourdan. In: Spitz, F., Janeau, G., Gonzalez, G., Anlagnier, S. (eds). *Ongules/Ungulates 91 Toulouse, France 1991*, 513–518.

Barandun, J. (1983). Afforestation at high altitudes. *Schweizerische Zeitschrift fur Forstwesen*, **134**, 431–441.

Beals, E.W., Cottam, G.W. and Vogal, R.J. (1960). Influence of deer on vegetation of the apostle islands, Wisconsin. *Journal of Wildlife Management*, **24**, 68–79.

Bergstrom, R. and Danell, K. (1987). Effects of simulated winter browsing by moose on morphology and biomass of two birch species. *Journal of Ecology*, **75**, 533–544.

Bideau, E., Vincent, J.P., Gerard, J.F. and Maublanc, M.L. (1992). Influence of sex and age on space occupation by roe deer (*Capreolus capreolus* L.). In: Spitz, F., Janeau, G., Gonzalez, G., Anlagnier, S. (eds). *Ongules/Ungulates 91 Toulouse, France 1991*, 263–266.

Brandner, T.A., Peterson, R.O. and Risenhoover, K.L. (1990). Balsam fir on isle royale — effects of moose herbivory and population density. *Ecology*, **71**, 155–164.

Cibien, C., Boutin, J.M. and Maizeret, C. (1988). Impact of roe deer (*Capreolus capreolus*) on vegetation in relation to population density and type of woodland. *Zeitschrift Jagdwissenschaft*, **34**, 232–241.

Corbet, G.B. and Harris, S. (eds) (1991). *The Handbook of British Mammals*. Blackwells, Oxford, 588pp.

Crawley, M.J. (1989). The relative importance of vertebrate and invertebrate herbivores in plant population dynamics. In: Bernays, E.A. (ed.). *Insect–Plant Interactions*. CRC Press. Boca Raton, Florida, 45–71.

Crawley, M.J. (1990). The population dynamics of plants. *Philosophical Transactions of the Royal Society, London, Ser. B,* **330**, 125–140.

Crouch, G.L. (1966). Effects of simulated deer browsing on Douglas fir seedlings. *Journal of Forestry*, **64**, 322–326.

Danilkin, A.A. (1992). Population structure. In: Sokolov, V. E. (ed.). *European and Siberian Roe Deer*. Nauka, Moscow, 160–184.

Denis M. (1992). Some data on a high density population of field roe deer. In: Spitz, F., Janeau, G., Gonzalez, G., Anlagnier, S. (eds). *Ongules/Ungulates 91 Toulouse, France 1991*, 519–521.

Dennis, R.L.H. (ed.) (1992). *The Ecology of Butterflies in Britain.* Oxford University Press, Oxford.

Don, B.A.C. (1981). Spatial dynamics and individual quality in a population of the grey squirrel, *Sciurus carolinensis* Gmelin. DPhil Thesis, Oxford University.

Eiberle, K. and Bucher, H. (1989). Interdependence between browsing of different tree species. *Zeitschrift für Jagdwissenschaft*, **35**, 235–244.

Eiberle, K. and Nigg, H. (1987). Basis for assessing game browsing in montane forests. *Schweizerische Zeitschrift fur Forstwesen*, **138**, 747–785.

Eiberle, Von K. (1975). Result of simulation of game damage through shoot cutting. *Schweizerische Zeitschrift fur Forstwesen*, **126**, 821–838.

Fenner, M. (1991). Irregular seed crops in forest trees. *Quarterly Journal of Forestry*, **85**, 166–172.

Fitzgibbon, C.D. (1993). The distribution of grey squirrel dreys in farm woodland: the influence of wood area, isolation and management. *Journal of Applied Ecology*, **30**, 736–742.

Frelich, L.E. and Lorimer, C.G. (1985). Current and predicted long-term effects of deer browsing in hemlock forests in Michigan, USA. *Biological Conservation*, **34**, 99–120.

Fuller, R.J. (1982). *Bird Habitats in Britain.* Poyser, Calton.

Fuller, R.M. and Boorman, L.A. (1977). The spread and development of *Rhododendron ponticum* L. on dunes at Winterton, Norfolk, in comparison with invasion by *Hippophae rhamnoides* L. at Saltfleetby, Lincolnshire. *Biological Conservation*, **12**, 83–94.

Furrh, P.L. and Ezell, A.W. (1982). Pine utilisation by deer in young Loblolly pine plantations in east Texas. *Proceedings of the Second Biennial Southern Silvicultural Research Conference*, Atlanta, Georgia, November 4–5, 1982, USDA Forest Service, Southern Forest Experiment Station, General Technical Report SE 24, 1983, 496–503.

Gill, R.M.A. (1990). *Monitoring the Status of European and North American Cervids.* GEMS Information Series Global Environment Monitoring System, United Nations Environment Programme., Nairobi, Kenya, 277pp.

Gill, R.M.A. (1992a). A review of damage by mammals in north temperate forests: 1. Deer. *Forestry*, **65**, 145–169.

Gill, R.M.A. (1992b). A review of damage by mammals in north temperate forests: 2. Small mammals. *Forestry*, **65**, 281–308.

Gill, R.M.A. (1992c). A review of damage by mammals in north temperate forests: 3. Impact on trees and forests. *Forestry*, **65**, 363–388.

Gladfelter, L. (1984). Midwest farmbelt. In: Halls, L.K. (ed.). *White-tailed Deer.* Stackpole Books, Pennsylvania.

Gregory, S.C. (1986). The development of stain in wounded Sitka spruce stems. *Forestry*, **59**, 199–208.

Gurnell, J. (1981). Woodland rodents and tree seed supplies. In: Chapman, J.A. and Pursley, D. (eds). *Worldwide Furbearer Conference Proceedings, Virginia, USA, 1981*, 1191–1214.

Gurnell, J. (1983). Squirrel numbers and the abundance of tree seeds. *Mammal Review*, **13**, 133–148.

Gurnell, J. (1987). *The Natural History of Squirrels.* Christopher Helm, London, 201pp.

Gurnell, J. (1989). Demographic implications for the control of grey squirrels. In: Putman, R.J. (ed.). *Mammals as Pests.* Chapman and Hall, London, 131–143.

Gurnell, J. (1991). The grey squirrel. In: Corbet, G.B. and Harris, S. (eds). *The Handbook of British Mammals.* Blackwell, Oxford, 186–190.

Gurnell, J. (1993). Tree seed production and food conditions for rodents in an oak wood in southern England. *Forestry*, **66**, 291–315.

Gurnell, J. and Pepper, H. (1989). Perspectives in the management of red and grey squirrels. In: Jardine, D.C. (ed.). *Wildlife Management in Forests, Lancaster 1987.* Institute of Chartered Foresters, 92–109.

Gurnell, J. and Pepper, H. (1993). A critical look at conserving the British red squirrel *Sciurus vulgaris. Mammal Review*, **23**, 127–137.

Haber, A. and Matuszewski, G. (1966). Observations on wild rabbits (*Oryctolagus cuniculus*) in forest areas of Poland. *Zeszyty Naukowe Szkoly Glownej Gospodarstwa wiejskiego w Warszawie*, **9**, 87–101.

Halls, L.K. (1984). *White-tailed Deer*. Stackpole Books, Pennsylvania, USA.

Halls, L.K. and Alcaniz, R. (1968). Browse plants yield best in forest openings. *Journal of Wildlife Management*, **32**, 185–186.

Hampshire, R. (1985). A study on the social and reproductive behaviour of captive grey squirrels *Sciurus carolinensis*. PhD thesis, University of Reading.

Heikkila, R. (1992). Moose browsing in a Scots pine plantation mixed with deciduous tree species. *Acta Forestalia Fennica*, **224**, 1–13.

Heikkila, R., Mikkonen, T. (1992). The effects of density of young Scots pine (*Pinus sylvestris*) stand on moose (*Alces alces*) browsing. *Acta Forestalia Fennica*, **231**, 1–14.

Henry, B.A.M. (1981). Distribution patterns of roe deer (*Capreolus capreolus*) related to the availability of food and cover. *Journal of Zoology, London*, **194**, 271–275.

Hill, D.A., Lambton, S., Proctor, I. and Bullock, I. (1991). Winter bird communities in woodland in the Forest of Dean, England, and some implications of livestock grazing. *Bird Study*, **38**, 57–70.

Hofmann, R.R. (1985). Digestive physiology of the deer — their morphophysiological specialisation and adaptation. In: *The Biology of Deer Production*. Dunedin, New Zealand, 13–18 February 1983. Bulletin 22. The Royal Society of New Zealand, 393–408.

Holisova, V., Obrtel, R., Kozena, I. and Danilkin, A.A. (1992). Feeding. In: Sokolov, V. E. (ed.). *European and Siberian Roe Deer*. Nauka, Moscow, 124–139.

Holloway, C.W. (1967). The effect of red deer and other animals on naturally regenerated Scots pine. PhD thesis, University of Aberdeen.

Horsley, S.B. and Marquis, D.A. (1984). Interference by weeds and deer with Allegheny hardwood reproduction. *Canadian Journal of Forest Research*, **13**, 61–69.

Huss, J. and Olberg-Kalfass, R. (1982). Unwanted interactions between weed control treatments and roe deer damage in Norway spruce plantations. *Allgemeine Forstzeitschrift*, **74**, 1329–1331.

Jensen, T.S. (1985). Seed–predator interactions of European beech, *Fagus sylvatica* and forest rodents, *Clethrionomys glareolus* and *Apodemus flavicollis*. *Oikos*, **44**, 149–156.

Johnson, T.H. (1984). Habitat and social organisation of roe deer (*Capreolus capreolus*). PhD thesis, University of Southampton.

Jones, S.B., deCalesta, D. and Chunko, S.E. (1993). White-tails are changing our woodlands. *American Forests*, **99**, 20–54.

Kaluzinski, J. (1982). Dynamics and structure of a field roe deer populations. *Acta Theriologica*, **27**, 385–408.

Kammerlander, H. (1978). Structure and regeneration of 'Forets Jardinees' in Kufstein (Tirol) and their vulnerability to browsing. *Schweizerische Zeitschrift fur Forstwesen*, **129**, 711–726.

Kay, S. (1993). Factors affecting severity of deer browsing damage within coppiced woodlands in the south of England. *Biological Conservation*, **63**, 217–222.

Kennedy, C.E.J. and Southwood, T.R.E. (1984). The number of species of insects associated with British trees: A re-analysis. *Journal of Animal Ecology*, **53**, 455–478.

Kennedy, T.H. (1956). A note on myxomatosis and natural regeneration. *Scottish Forestry*, **10**, 112.

Kenward, R.E. (1983). The causes of damage by red and grey squirrels. *Mammal Review*, **13**, 159–166.

Kenward, R.E. (1985). Ranging behaviour and population dynamics in grey squirrels. In: Sibly, R. M. and Smith, R.H. (eds). *Behavioural Ecology: Ecological Consequences Of Adaptive Behaviour*. Blackwell, Oxford, 319–330.

Kenward, R.E. and Parish, T. (1986). Bark-stripping by grey squirrels (*Sciurus carolinensis*). *Journal of Zoology*, **210**, 473–481.

Kenward, R.E., Parish, T. and Doyle, F. (1988). Grey-squirrel bark stripping. II Management of woodland habitats. *Quarterly Journal of Forestry*, **82**, 87–94.

Kraus, P. (1987). The use of vegetation by red deer as an indicator of their population density. *Zeitschrift Jagdwissenschaft*, **33**, 42–59.

Krefting, L.W. and Stoekeler, J.H. (1953). Effect of simulated snowshoe hare and deer damage on planted conifers in the Lake States. *Journal of Wildlife Management*, **17**, 487–494.

Leopold, A., Sowls, L.K. and Spencer, D.L. (1947). A survey of over-populated deer ranges in the United States. *Journal of Wildlife Management*, **11**, 162–177.

Lewis, C. (1980). Simulated cattle injury to planted slash pine: defoliation. *Journal of Range Management*, **33**, 337–340.

Lloyd, H.G. (1970). Post myxomatosis rabbit populations in England and Wales. *EPPO Public Series. A*, 197–215.

Lloyd, H.G. (1983). Past and present distribution of red and grey squirrels. *Mammal Review*, **13**, 69–80.

Loffler, H. (1975). The spreading of wound rot in Norway spruce. *Forstwissenschaft Centralblatt*, **94**, 175–183.

Loudon, A.S.I. (1987). The Influence Of Forest Habitat Structure On Growth Body Size And Reproduction. In: Wemmer, C. M. (ed.). *Roe Deer Capreolus-Capreolus L. Biology and Management of the Cervidae Front Royal, Virginia, USA*. Smithsonian Institution Press, 577. USA.

Lyly, O. and Saksa, T. (1992). The effect of stand density on moose damage in young *Pinus sylvestris* stands. *Scandinavian Journal of Forest Research*, 7, 393–403.

MacDonald, J.M. and Lockhart, S.F.M. (1953). Some early observations on the natural regeneration of conifers in Scotland. *Scottish Forestry*, 7, 79–85.

Mackinnon, K. (1976). Home range, feeding ecology and social behaviour of the grey squirrel (*Sciurus carolinensis* Gmelin). PhD thesis, University of Oxford.

Maizeret, C. and Ballon, P. (1990). Analysis of causal factors behind cervid damage on the cluster pine in the landes of Gascony. *Gibier Faune Sauvage*, 7, 275–291.

Marquis, D.A. and Brenneman, R. (1981). The Impact of Deer on Forest Vegetation in Pennsylvania. General Technical Report NE-65. USDA Forest Service, 1–8.

McIntyre, E.B. (1975). Bark stripping by ungulates. PhD thesis, University of Edinburgh.

Miles, J. (1972). Experimental establishment of seedlings on a southern English heath. *Journal of Ecology*, **60**, 225–234.

Miller, G.R., Kinnaird, J.W. and Cummins, R.P. (1982). Liability of saplings to browsing on a red deer range in the Scottish Highlands. *Journal of Applied Ecology*, **198**, 941–951.

Mitchell, B., Staines, B.W. and Welch, D. (1977). *Ecology of Red Deer: A Research Review Relevant to Their Management in Scotland*. Natural Environment Research Council, Institute of Terrestrial Ecology, Cambridge.

Mitchie, C.Y. (1868). Report on trees not liable to be eaten by rabbits. *Transactions of the Highland and Agricultural Society of Scotland*, 446–456.

Monk, K. (1986). Food selection and feeding behaviour in farmland rabbits. MAFF/University of Reading research report.

Newsome, A. (1990). The control of vertebrate pests by vertebrate predators. *Trends in Ecology and Evolution*, **5**, 187–191.

Nielsen, B.O. (1977). Beech seeds as an ecosystem component. *Oikos*, **29**, 268–274.

Nilsson, S.G. (1985). Ecological and evolutionary interactions between reproduction in beech *Fagus sylvatica* and seed eating animals. *Oikos*, **44**, 157–164.

Ormrod, P.C. (1991). Grey squirrels v Spanish chestnut. *Quarterly Journal of Forestry*, **85**, 272.

Papageorgiou, N.K. (1978). Use of forest openings by roe deer as shown by pellet group counts. *Journal of Wildlife Management*, **42**, 650–654.

Pawsey, R.G., Gladman, R.J. (1965). *Decay in Standing Conifers Developing from Extraction Damage*. Forest Record. Forestry Commission, Edinburgh.

Pepper, H.W. (1977). Protection of trees, shrubs, and garden plants from damage by deer. *Deer*, 4, 150–152.

Pepper, H.W. (1978). *Chemical Repellants*. Forestry Commission Leaflet 73, 8pp.

Pepper, H.W. (1989a). *Grey Squirrels and the Law*. Forestry Commission Research Information Note 191, Forestry Commission, Farnham.

Pepper, H.W. (1989b). *Hopper Modification for Grey Squirrel Control*. Forestry Commission Research Information Note 153, Forestry Commission, Farnham.

Pepper, H.W. (1990). *Grey Squirrels Damage Control With Warfarin*. Forestry Commission Research Information Note 180, Forestry Commission, Farnham.

Pepper, H.W. (1992). *Forest Fencing*. Forestry Commission Bulletin 102. HMSO, London.

Peterken, G.F. and Jones, E.W. (1989). Forty years of change in Lady Park Wood: The young-growth stands. *Journal of Ecology*, 77, 401–429.

Picard, J.F. (1976). Feeding preferences of deer and their consequences: first conclusions from two years' experimentation. *Revue Forestiere Francaise*, **28**, 106–114.

Piechocki, R. (1994). Who's afraid of the wandering wolf? *New Scientist*, 2 April 1994, 19–21.

Pigott, C.D. (1983). Regeneration of oak-birch woodland following exclusion of sheep. *Journal of Ecology*, **71**, 629–646.

Pigott, C.D. (1985). Selective damage to tree seedlings by bank voles. *Oecologia*, **67**, 367–371.

Pigott, C., Newton A. and Zammit, S. (1991). Predation of acorns and oak seedlings by grey squirrels. *Quarterly Journal of Forestry*, **85**, 173–178.

Potter, M.J. (1991). *Treeshelters*. Forestry Commission Handbook 7. HMSO, London. 48pp.

Prior, R. (1968). *The Roe Deer of Cranbourne Chase*. Oxford University Press, London.

Putman, R.J. (1994). Deer damage in coppice woodlands: An analysis of factors affecting the severity of damage and options for management. *Quarterly Journal of Forestry*, **88**, 45–54.

Putman, R.J., Edwards, P.J., Mann, J.E.E., Howe, R.C. and Hill, S.D. (1989). Vegetational and faunal change in an area of heavily grazed woodland following relief of grazing. *Biological Conservation*, **47**, 13–32.

Rackham, O. (1975). *Hayley Wood: Its History and Ecology*. Cambridgeshire and Isle of Ely Naturalists Trust.

Ratcliffe, P.R. (1984). Population density and reproduction of red deer in Scottish commercial forests. *Acta Zoologica Fennica*, **172**, 191–192.

Ratcliffe, P.R. (1987). The management of red deer in the commercial forests of Scotland related to population dynamics and habitat changes. PhD thesis, University of London.

Ratcliffe, P.R. and Mayle, B.A. (1992). *Roe Deer Biology and Management*. Forestry Commission Bulletin 105. HMSO, London, 36pp.

Rees, W.A., Ross, J., Cowan, D.P., Tittensor, A.M. and Trout, R.C. (1985). Human control of rabbits. In: *Humane Control of Land Mammals and Birds*. UFAW Symposium, Guildford, 1984, 96–103.

Repo, S. and Loyttyniemi, K. (1985). The effect of immediate environment on moose (*Alces alces*) damage in young Scots pine plantations. *Folia Forestalia*, **626**.

Rousi, M. (1990). Breeding forest trees for resistance to mammalian herbivores — a study based on European white birch. *Acta Forestalia Fennica*, **210**, 1–20.

Rowe, J. (1982). Roe research in relation to management in British woodlands. *Roe and Red deer in British Forestry*. June 1982. The British Deer Society and Forestry Commission, 25–41.

Rowe, J. and Gill, R. (1985). The susceptibility of tree species to damage by grey squirrels in England and Wales. *Quarterly Journal of Forestry*, **79**, 183–190.

Rowe, J.J. (1984). Grey squirrel bark-stripping damage to broadleaved trees in southern Britain up to 1983. *Quarterly Journal of Forestry*, **78**, 231–236.

Sempere, A., Garreau, J. and Boissin, J. (1980). Seasonal variations in territorial marking activity and testosterone in adult male roe deer. *Compte Rendu Academi des Sciences de Paris Series D*, **803**, 803–806.

Shaw M.W. (1968). Factors affecting the natural regeneration of sessile oak (*Quercus petraea*) in North Wales. 1. A preliminary study of acorn production, viability and losses. *Journal of Ecology*, **56**, 565–583.

Sheail, J. (1971) *Rabbits and Their History*. David and Charles, Newton Abbot.

Shorten, M. (1954). *Squirrels*. Collins, London.

Shorten, M. (1957). Damage caused by squirrels in Forestry Commission areas, 1954–6. *Forestry*, **30**, 151–172.

Silvertown, J.W. (1980). The evolutionary ecology of mast seeding in trees. *Biological Journal of the Linnean Society*, **14**, 235–250.

Skogland, T. (1991). What are the effects of predators on large ungulate populations. *Oikos*, **61**, 401–411.

Smith, C.C. and Follmer, D. (1972). Food preferences of squirrels. *Ecology*, **53**, 82–91.

Smith, C.J. (1980). *Ecology of the English Chalk*. Academic Press, London.

Smith, J.J. (1954). Rabbit clearance in the King's forest. *Journal of the Forestry Commission*, **23**, 70–72.

Staines, B.W. and Ratcliffe, P.R. (1987). Estimating the abundance of red (*Cervus elaphus*) and roe (*Capreolus capreolus*) deer and their current status in Great Britain. *Symposium of the Zoological Society of London*, **58**, 131–152.

Staines, B.W. and Welch, D. (1984). Habitat selection and impact of red deer and roe deer in a Sitka spruce plantation. *Proceedings of the Royal Society of Edinburgh*, **82**, 303–319.

Stapanian, M.A. and Smith, C.C. (1986). How fox squirrels influence the invasion of prairies by nut-bearing trees. *Journal of Mammalogy*, **67**, 326–332.

Stehle, K. (1986). Silviculture and roe deer: A successful synthesis on a private estate. *Allgemeine Forstzeitschrift*, **49**, 1224–1227.

Stewart, G.H. and Burrows, L.E. (1989). The impact of white-tailed deer *Odocoileus virginianus* on regeneration in the Coastal Forests of Stewart-Island, New-Zealand. *Biological Conservation*, **49**, 275–293.

Sumption, K.J. and Flowerdew, J.R. (1985). The ecological effects of the decline in rabbits (*Oryctolagus cuniculus* L.) due to myxomatosis. *Mammal Review*, **15**, 151–186.

Tansley, A.G. (1949). *The British Islands and Their Vegetation.* Cambridge University Press, Cambridge.

Tanton, M.T. (1965). Acorn destruction potential of small mammals and birds in British Woodlands. *Quarterly Journal of Forestry*, **3**, 230–234.

Taylor, J.C. (1969). Social structure and behaviour in a grey squirrel population. PhD thesis, University of London.

Thirgood S.J. and Staines B.W. (1989). Summer use of young stands of restocked by red and roe deer. *Scottish Forestry*, **43**, 183–191.

Thomas, A.S. (1960). Changes in vegetation since the advent of myxomatosis. *Journal of Ecology*, **48**, 287–306.

Thomas, A.S. (1963). Further changes in vegetation since the advent of myxomatosis. *Journal of Ecology*, **51**, 151–186.

Thompson, D.C. and Thompson, P.S. (1980). Food habits and caching behaviour of urban grey squirrels. *Canadian Journal of Zoology*, **58**, 701–710.

Thompson, H.V. and King, K.M. (1944). *The European Rabbit.* Oxford Science Publications, Oxford.

Thompson, H.V. and Wordon (1966). *The Rabbit.* Collins, London.

Thompson, I.D. (1988). Moose damage to pre-commercially thinned balsam fir stands in Newfoundland. *Alces*, **24**, 56–61.

Tilghman, N.G. (1989). Impacts of white-tailed deer on forest regeneration in Northwestern Pennsylvania. *Journal of Wildlife Management*, **53**, 524–532.

Tomback, D.F. (1982). Dispersal of whitebark pine seeds by Clarke's nutcracker: a mutualism hypothesis. *Journal of Animal Ecology*, **51**, 451–467.

Trout, R.C. and MacVicker, H.M. (1993). Making rabbits and small herbivores dislike young trees. XXIst IUGB conference, Halifax, 1993.

Trout, R.C., Ross, J., Tittensor, A.M. and Fox, A.P. (1992). The effect on a British wild rabbit population (*Oryctolagus cuniculus*) of manipulating myxomatosis. *Journal of Applied Ecology*, **29**, 679–686.

Trout, R.C., Tapper, S.C. and Harradine, J. (1986). Recent trends in the rabbit population in Britain. *Mammal Review*, **16**, 117–123.

Trout, R.C. and Tittensor, A.M. (1989). Can predators regulate wild rabbit *Oryctolagus cuniculus* L. density in England and Wales. *Mammal Review*, **19**, 153–173.

Trumbull, V.L., Zielinski, E.J. and Aharrah, E.C. (1989). The impact of deer browsing on the Allegheny forest type. *Northern Journal of Applied Forestry*, **6**, 162–165.

Turcek, F.J. and Kelso, L. (1968). Ecological aspects of food transporation and storage in the Corvidae. *Communications in Behavioural Ecology*, **1**, 277–297.

Warren, M. (1992). Britain's vanishing fritillaries. *British Wildlife*, **3**, 282–296.

Warren, M. and Thomas, J.A. (1992). Butterfly responses to coppicing. In: Buckley, G.P. (ed.). *Ecology and Management of Coppice Woodlands.* Chapman and Hall, London, 249–270.

Welch, D., Staines, B., Scott, D. and French, D. (1992). Leader browsing by red and roe deer on young Sitka spruce trees in western Scotland. 2. Effects on growth in tree form. *Forestry*, **65**, 309–330.

Welch, D., Staines, B.W., Catt, D.C. and Scott, D. (1990). Habitat usage by red (*Cervus elaphus*) and roe (*Capreolus capreolus*) deer in a Scottish Sitka spruce plantation. *Journal of Zoology*, **221**, 453–476.

Welch D., Staines, B.W., Scott, D. and Catt, D.C. (1987). Bark stripping damage by red deer in a Sitka spruce forest in western Scotland. I: Incidence. *Forestry*, **60**, 249–262.

Welch, D., Staines, B.W., Scott, D., French, D.D. and Catt, D.C. (1991). Leader browsing by red and roe deer on young Sitka spruce trees in Western Scotland. 1. Damage rates and the influence of habitat factors. *Forestry*, **64**, 61–82.

Whelan, C.J., Willson, M.F., Tuma, C.A. and Souza-Pinto, I. (1991). Spatial and temporal patterns of postdispersal seed predation. *Canadian Journal of Botany*, **69**, 428–436.

Williams, M. (1989). *Americans and Their Forests*. Cambridge University Press, Cambridge, 628pp.

Winter, T.G. (1983). *A Catalogue of Phytophagous Insects and Mites on Trees in Great Britain*. Forestry Commission Booklet 53. HMSO, London.

Worrell, R. and Nixon, C.J. (1991). *Factors Affecting the Natural Regeneration of Oak in Upland Britain — a Literature Review*. Forestry Commission Occasional Paper 31. Forestry Commisson, Edinburgh.

Zeltner, J. (1979). Impoverishment of tree species mixtures as a result of roe deer populations. *Schweizerische Zeitschrift fur Forstwesen*, **130**, 81–84.

14 Ecological Planning in New Woodlands

R. FERRIS-KAAN

INTRODUCTION

This chapter brings together the many themes taken up in previous chapters. It follows a chronosequence of the woodland creation and management process: setting objectives; site assessment and evaluation; site layout; establishment options and management considerations; and finally examining research needs. The creation of new woodland on a site presents numerous ecological challenges and opportunities and, in order to create a diverse, functioning woodland, to meet multiple objectives, it is important to take a long-term view of sustainable management.

Setting objectives

Agreed management objectives should be set out in a management plan and, if appropriate, offered for consultation. The plan should contain a list of intentions, prescriptions for achieving these, and a time-frame to the overall strategy. All of the data required may not be initially available and so the plan needs to be viewed as an evolving document. Initial work may need to be targeted on areas identified as presenting early opportunities for woodland creation (Marsh, 1993), or directed towards specific tasks with high likelihood of success. For example, the identification of optimal precursor vegetation, which indicates that the site may be well suited to future woodland (Rodwell and Patterson, 1994), can aid such targeting of resources.

SITE ASSESSMENT AND EVALUATION

Information must be gathered to form an ecological database, which enables sites to be evaluated and graded on their merits for a range of objectives. Information gathered will help in:

- Identification of factors influencing site layout, including the balance between woodland and open ground, and woodland shape (i.e. opportunities for habitat and landscape enhancement).
- Deciding the specification for tree establishment.

The Ecology of Woodland Creation. Edited by Richard Ferris-Kaan.

● Development of broad management strategies.
● Provision of baseline data against which any future changes can be monitored.

Broad habitat description

The first stage of the process should be to describe the site, attempting to evaluate both its existing and potential value, including landscape character and habitat types (see Bell, Chapter 3; Rodwell and Patterson, 1994). Habitat description is best based on dominant vegetation types according to a recognised system of vegetation classification such as the NVC (see Rodwell and Patterson, Chapter 5), some measure of their extent, and the location of key features. It will then be possible to identify areas of high potential for habitat enhancement or creation, and the likely importance of biogeographical factors (see Spellerberg, Chapter 4); for example, proximity to potential colonising sources (see Harmer and Kerr, Chapter 8), dispersal networks, and buffer zones.

Physico-chemical factors

Site physico-chemical characters will define limitations for new woodland:

● Soil (texture, profile thickness, density and chemistry)
● Climate (susceptibility to frost, rainfall, temperature)
● Topography (gradient, microrelief).

Soil type is of vital importance in determining tree species choice and the associated plant communities (see Rodwell and Patterson, Chapter 5; Moffat and Buckley, Chapter 6), and the establishment methods to be used (Moffat and Bending, 1992; Harmer and Kerr, Chapter 8). Climatic variables will also exert considerable influence over species choice and vegetation succession. Tree species differ in their susceptibility to frost, with Norway spruce (*Picea abies*) and common alder (*Alnus glutinosa*) being notably frost-tolerant. Rainfall has relatively little effect on the choice of broadleaved species, but is an important determinant of species choice for conifers. For example, good growth of Douglas fir (*Pseudotsuga menziesii*) on sheltered brown-earth sites in the west of Britain, and Corsican pine (*Pinus nigra* var. *maritima*) on similar sites in eastern Britain, is mainly due to rainfall (Hibberd, 1988). Climatic considerations overlap to some degree with topographical factors, and can affect patterns of natural regeneration or planting design. Some trees and shrubs found in old semi-natural woods (for example, beech (*Fagus sylvatica*), hornbeam (*Carpinus betulus*), holly (*Ilex aquifolium*)) tend only to colonise in the later stages of natural successions. Beech shows a sensitivity to frost which may present problems in open, new woods. Possible strategies are to delay the planting of these species until other trees are established, to provide suitable conditions for establishment, or await natural colonisation (Rodwell and Patterson, 1994). Investment in any of these options should be restricted to the most suitable parts of the site.

Biotic factors

The biological character of the site may influence management decisions in a number of ways:

- Likely future direction of successional processes
- Below-ground biotic processes
- Influences on choice and layout of tree species.

An assessment of the plant communities present can indicate the likely future direction of successional processes and indicate suitable tree species and the likelihood of successful establishment. Rodwell and Patterson (1994) provide a useful basis to this through the concepts of optimal precursor vegetation and desired invaders. Le Duc *et al.* (1992) have developed a method for predicting the probability of species occurring in a particular site type, using data from systematic surveys of existing species distributions, plus soil and land classification data. However, this system does not account for differences in dispersal and establishment ability, such that, for isolated woods, it would still be necessary to determine suitable slow colonisers, i.e. ancient woodland indicator species (Cunningham, 1993), by other means such as the NVC or a similar ecological database.

A soil assay may be used to assess the biotic conditions below ground, to provide an indication of the likely vegetation development on a site (see Harris and Hill, Chapter 7). Not only can the soil microbial communities present help to initiate and direct this succession, but the composition of buried propagules (i.e. the soil seed-bank) can influence the plant community types which will regenerate. These are unlikely to be made up of 'true woodland species', since few rely on seed for their propagation. The plant species forming small but persistent seedbanks (Thompson and Grime, 1979) tend to be those more typical of secondary, disturbed habitats.

Many of the sites on which new woodland is planned will be highly modified and the most effective way to ensure that the character and composition of local and regional patterns of woodland is maintained is to assess species composition of established secondary woods on broadly similar soil types. It is much more difficult to draw inferences based on local ancient woodland patterns due to the dissimilarity in soil condition (see Moffat and Buckley, Chapter 6). Small-scale variation in species abundance and distribution in established woods can also indicate how the location of planting of individual species and mixtures in new woods can be related to changes in shade, soil moisture and fertility across the site.

SITE LAYOUT

The landscape impact of new and even small-scale woodland can be great, requiring careful design at the site level. If the objective is to create large woods with true interior conditions, large, compact blocks must be planned or existing woodlands extended. Furthermore, to ensure good connectivity between woods, it is important to plan for a concentration of woods in certain areas. Peterken (in press) refers to this pattern of new woodland as multiple nucleus afforestation, with the nuclei focused on existing concentrations. There should be a preference for species recruitment through siting new woods adjacent to existing woods, by expanding initial new woods, and through the addition of new woods to existing linear concentrations such as hedgerows or riparian woods.

The design strategy must integrate the mix of habitats and zoned areas (to satisfy different demands) into a coherent whole (Forestry Authority, 1994; Marsh, 1993), through, for example, imaginative design of the scale and pattern of compartments within a woodland.

The balance between wooded and open ground

How much of the site should be wooded, and what is the best layout of wooded and open space? The current recommendation is that up to 20% of the total woodland area should be maintained as open habitats (Forestry Commission, 1990; Watkins, 1990). The more extensive the area for new woodland, the more likely it is to contain a mosaic of habitat types, supporting a diverse flora and fauna (see Spencer, Chapter 1). Small or numerous well-dispersed woods will have a much greater proportion of 'edge' habitat, and further reduction in wooded area by the inclusion and maintenance of extensive open ground may leave an insufficient woodland interior.

Open ground should be sited in habitats naturally less suited to woodland, such as wetland, crags, screes, shallow soils and exposed ridges (Rodwell and Patterson, 1994). These areas are usually of intrinsic nature conservation value, and add to the structural diversity and amenity value of the woodland. Adjacent planting must not jeopardize, and if possible should enhance, the value of these open habitats (Hibberd, 1988).

Most sites are entirely capable of supporting tree growth, but none the less areas should be identified for maintenance as open ground to meet nature conservation as well as amenity and recreation objectives. Site assessment should have identified vulnerable areas of semi-natural vegetation which should form the basis of an open space network around which the potential of other necessarily unwooded areas such as footpaths, access and extraction routes, or wayleaves can be developed (Marsh, 1993). By identifying open space networks early on, rather than creating them at a later stage when the trees have grown, greater stability in vegetation pattern and process is ensured. This network of permanent open space forms the matrix onto which temporary open areas can be added during subsequent management operations.

Open space and variable tree spacing around the edge of new woodland help to integrate the wood more naturally into the landscape, from both an ecological and visual perspective (Rodwell and Patterson, 1994). These may also act as ecological buffer zones, minimising any negative impacts from surrounding land-uses (see Spellerberg, Chapter 4).

Shape of new woodland

The ideal shape of nature reserves has been long debated (see Spellerberg, Chapter 4), and this debate may be applied to new woods. If, for example, new woods are to be established some distance away from existing woodland, then most plant and animal species must be acquired by colonisation. In this case, a shape to maximise the probability of colonisation is desirable. In order to make the most contacts with other habitat blocks, from which flora and fauna may be 'trawled', linear shapes may be optimal (Peterken, 1994). In well-colonised woodlands, however, the main concern is that existing species are 'retained'. This may require more compact woods to ensure woodland interior conditions. Consequently, the optimum shape required to satisfy both major objectives is a compact wood with arms radiating outwards in several directions in good contact with linear habitats such as hedgerows and riparian corridors.

ESTABLISHMENT OPTIONS

The question of whether to plant trees or rely on natural colonisation is addressed by Harmer and Kerr in Chapter 8. Seldom is the choice clear cut as many factors influence the decision. The issue can be extended beyond the tree component of the new woodland, to include the introduction of sub-canopy vegetation and even the translocation of less mobile fauna such as many deadwood invertebrates.

Influences on management

Site factors, already alluded to, are the most significant influences, although ecological considerations, time and money will also influence decisions. The relative importance of these factors is dependent on the hierarchy of objectives set, and will greatly influence management options (e.g. the amount of time and money allocated to ground flora development, or the need for a commercial return from the tree crop).

Site capture by tree planting

There may be good social reasons why the matrix of woodland trees needs to be rapidly established. In urban areas the appearance of a woodland framework is crucial to demonstrate that a woodland is being created on the site. If the site is degraded, the establishment of a tree canopy may play a vital role in reclamation (see Moffat and Buckley, Chapter 6).

There are additional ecological reasons for ensuring a speedy capture of the site to woodland, such as to help the successful introduction and establishment of woodland ground flora. Cunningham (1993) has identified three main factors of importance in providing the necessary conditions for understorey development in new woods:

- Light intensity beneath the canopy
- Soil and litter conditions
- Biotic factors, especially competition.

In order to ensure a quick capture of a site to woodland, relatively close tree spacing is needed. Harmer and Kerr (Chapter 8) discuss spacing in greater detail, making the point that site conditions should influence the precise spacing adopted. It is, however, not advisable from a conservation point of view to adopt close spacing over the whole site, as there are a number of advantages to low stocking densities (Soutar, 1991):

- It is cheaper to plant and protect fewer trees and shrubs.
- Using fewer plants, it will be easier to find, collect and propagate local genetic stock.
- More room is left for future natural colonisation.
- There are much greater opportunities for the survival and development of ground flora species (where plant communities already present are of conservation importance).

Rodwell and Patterson (1994) propose a combination of planted clumps and open areas, subject to variations in site characteristics, to increase diversity in new woods. Species composition, size, location and distance between clumps can be varied to increase diversity and provide opportunity for natural colonisation. Quick-growing pioneer species such as birch (*Betula* spp.), willow (*Salix* spp.), alder (*Alnus* spp.), and aspen (*Populus tremula*) can be used as a matrix to provide protection and appropriate light conditions.

Natural colonisation

Ecologically, a reliance on some degree of natural colonisation is preferable to total reliance on planting, resulting in a more natural matching of trees and shrubs to the local conditions. This may produce a more irregular structure and natural appearance, and may offer direct benefits for wildlife. For example, there is evidence that woodland bird communities prefer distinct clumps of broadleaved trees (Bibby *et al.*, 1989).

Where natural succession is allowed to proceed unhindered, the course of development even on similar sites is extremely variable, and is affected by chance colonisation events and essentially random conditions at the time of colonisation (Peterken, 1994). Observations suggest that the formation of closed canopy woodland by natural colonisation takes much longer than with planting. An important consideration is which species are likely to become established most quickly, since these may determine the course of later succession, sometimes arresting the process. This is a form of succession referred to as 'inhibition' (Connell and Slatyer, 1977) or 'ecological inertia'. Consequently, the effects of short-term conditions and chance patterns in initial colonisation can be extremely long lasting.

If a high density of early colonisers becomes established, further regeneration niches (Grubb, 1977) may become very restricted for other species. As noted by Buckley and Knight (1989), the growth of such species may be too vigorous for climax species such as oak (*Quercus* spp.), beech, hornbeam and lime (*Tilia* spp.), necessitating some subsequent intervention. Pioneer woodland may be unattractive to the public, featuring dense thickets of woody vegetation, and will require management, such as thinning or enrichment planting, to become more visually appealing. A balance needs to be struck between planting and the encouragement of natural colonisation. In a survey carried out around four of the new Community Forest areas in England, natural colonisation had only reached an acceptable woodland standard on one fifth of vacant land, was patchy and tended to be species poor (S. J. Hodge, personal communication). Species with light, wind-dispersed seed (mainly birches and willows) tend to be over represented. Planting of bird and animal dispersed species (e.g. oak and hawthorn (*Crataegus* spp.)) and those with heavy wind-dispersed seed (e.g. ash (*Fraxinus excelsior*) and field maple (*Acer campestre*)) may be necessary in order to create a more natural species mix.

MANAGEMENT CONSIDERATIONS

Both site layout and the establishment methods influence, and in some cases dictate, subsequent management. This should be anticipated and considered at the design stage (Marsh, 1993). For example, where site capture has involved close spacing of the trees, it may be necessary to carry out early and more frequent thinnings, with the sole objective of maintaining an optimum light regime conducive to maximum sub-canopy diversity (Buckley and Knight, 1989). The formation of a good shrub layer will often require a thinning intensity of at least 10% greater than under conventional forestry practice (Anderson, 1989), with an early first thinning. The planting of complicated mixtures of trees and shrubs, often in 'intimate' mixes, usually necessitates active intervention to remove aggressive or superfluous individuals from an early stage in the development of the woodland (Helliwell, 1993). The design of the thinning regime to encourage sub-canopy vegetation will depend upon the specific type of vegetation desired and, more importantly, the existing on-site vegetation.

Vegetation patterns and responses

Using a classification system for describing vegetation patterns, it is often possible to identify plant species which are potential aggressors under certain conditions (see Rodwell and Patterson, Chapter 5). This approach can be utilised for both the canopy and sub-canopy component of a new wood, and can help predict the outcome of particular management operations. Autecological knowledge of the regenerative strategies of plant species (Grime et al., 1988) provides a valuable insight into the likely responses of species to disturbance, nutrient enrichment or stress due to lack of light, moisture or nutrients.

True woodland plants (as opposed to those of the woodland edge) cannot be established before there is sufficient shade to limit competition from light-demanding species, and conditions of relatively high humidity. Canopy management should aim to provide the preferred species with the optimum conditions to encourage their establishment and spread. Optimum soil conditions may be provided through a reduction in soil fertility on ex-arable sites, selecting species appropriate to site drainage and soil pH, and by hastening soil development towards woodland conditions, e.g. litter development, or an appropriate microbial community (see Harris and Hill, Chapter 7).

Effective measures for modifying the site to facilitate herbaceous colonisation include: cropping without fertilising before woodland establishment, soil movement prior to planting, in an attempt to restore the natural levels of site heterogeneity (Buckley and Knight, 1989); ploughing through land drains to reduce drainage and recreate a mosaic of wetter and drier sites; and allowing watercourses to become blocked (subject to agreement with the National Rivers Authority, or corresponding body outside of the UK, where necessary) to allow low-lying ground to flood or become moist (Peterken, 1994). Management objectives with regard to biotic factors should include reducing weed competition prior to any introduction (e.g. through selective use of herbicides, or by cultivation), minimising grazing and browsing during introduction, controlling shade-tolerant weed species after introduction, thereby optimising the survival of introduced species (Cunningham, 1993). Control measures at this stage require a high degree of selectivity, and careful monitoring is an essential component.

Using thinning as an example, it has been found that heavy thinning on highly fertile arable sites leads to the vigorous growth of arable weeds. These may then have the opportunity to recharge their seedbanks after thinning operations, so perpetuating the problem. Some arable weeds may have a value for some forms of wildlife, creating colourful glades which are attractive to birds and invertebrates (see Fuller et al., Chapter 11; Key, Chapter 10), but if a more shade-tolerant sub-canopy vegetation is sought, a regular but lighter selective thinning regime may be preferable, as this will allow a range of moderately shade-tolerant plants to flourish while restricting the dominance of the more competitive open ground plants.

Animal communities and herbivory

The likely development of animal communities in new woods can be predicted from factors at the landscape and site scale, as influenced by the establishment techniques used. Some form of intervention may be required to control population levels of certain pest species, both invertebrates and vertebrates. Gill *et al.* (Chapter 13) examine the impact that vertebrate herbivores, particularly deer, may have on the development of new woods. The ability to predict more accurately and assess populations of such species will allow proactive management strategies to be adopted which take account of the problem.

There is potential for conflict between management objectives. For example, a reliance on natural colonisation results in much greater quantities of accessible browse material than would be available under closed, shaded conditions (see Gill *et al.*, Chapter 13). This can lead to increasing roe deer (*Capreolus capreolus*) populations and hence increasing impact on ground vegetation. Furthermore, if new woods are designed to incorporate more edge habitat, then this has been shown to increase utilisation by both red deer (*Cervus elaphus*) and roe deer (Welch *et al.*, 1990). Biogeographical influences are also important, as new woods in agricultural habitats tend to be small and scattered, with a correspondingly large edge to area ratio, which provide cover and the opportunity to utilise neighbouring agricultural crops.

Habitat management can be used as a tool for reducing the risk of damage, such as selecting tree species for planting which are less susceptible to damage or by choosing a site that is not immediately adjacent to favourable habitat. At the outset, woodland design needs to take account of the potential impact of pest species, but where other objectives have to take precedence, subsequent management must contain the impact. Planting, colonisation or regeneration of vulnerable species will often necessitate the use of fencing or individual tree protection. Similar considerations are likely to apply when coppicing is carried out, especially where coupes are enclosed by suitable areas of cover for deer. Woodland edge shrub belts may be severely browsed by deer and other vertebrate herbivores, thereby reducing valuable structural and species diversity (Thornber, 1993; Buckley, 1994). In order to prevent a loss of wildlife value, particularly for invertebrates and songbirds, these coppice belts may need to be protected following management.

The visible presence of wildlife in woods, particularly large herbivores, is a valuable amenity asset. This may present some conflict with silvicultural and conservation objectives; for example, the presence of deer above a particular density can limit or prevent natural regeneration and influence the relative abundance of key plant species for other animal groups. In these situations, clear prioritisation of objectives will indicate which species must be compromised.

Monitoring change

Woodland creation is unpredictable, particularly on man-modified sites. The direction of vegetation succession and the eventual pattern of plant communities following species introductions depends on many stochastic factors (Cunningham, 1993). For example, edaphic conditions may change, leading to replacement of one species by another, since the micro-environment for regeneration and establishment of introduced species is likely to be critical. In view of this, the benefits from monitoring of changes in response to management in developing woodlands are likely to be even greater than in long-established woods.

Monitoring should be an integral part of woodland management (Ferris-Kaan and Patterson, 1992) and should be written into management plans at the outset, with clear objectives identified. To be effective, monitoring must be consistently repeatable, and the protocol, its costing and need for staff expertise should be well-planned. For particular taxonomic groups, such as flowering plants, butterflies and songbirds, there are likely to be many suitably experienced amateur naturalists able to help with field assessments.

CURRENT AND FUTURE RESEARCH

Monitoring vegetation composition is the most urgent priority in many secondary woods, to allow comparison with that of ancient woodland and, drawing on knowledge of regional species distributions, to lay down a baseline for what may be possible in a new woodland site. Cunningham (1993) suggests that the outcome of different planting patterns and densities needs to be investigated, along with long-term monitoring of natural colonisation in secondary woodland. Secondary woods which have regenerated on abandoned arable land and the urban fringe offer an indication of the likely direction of the development of new woods on such sites. Experience from old field sites in the United States, in particular from experimental work in New Jersey (Myster, 1993), has highlighted the critical importance of 'regeneration windows' and has suggested that the net effect of influences on seed and seedling processes, particularly predation, is to produce very low probabilities of establishment (Gill and Marks, 1991).

Succession towards woodland tends to be rapid at first, but then becomes progressively slower. Peterken and Game (1984), studying the influence of historical factors on the number and distribution of vascular plant species in Lincolnshire woods, found that secondary woods which formed between 200 and 300 years ago were no richer than those which formed in the last 100 years. Further monitoring of other biotic factors needs to be considered, such as the impact of grazing and browsing (see Gill *et al.*, Chapter 13), and rates of colonisation by less mobile species groups such as certain invertebrates and soil borne organisms. Given the likely proximity of new woods to towns, there is a need to monitor the impact of recreational use on various components of the developing woodland ecosystem. This is a field of research which is developing rapidly, following early studies in urban fringe woodlands in the Netherlands (e.g. van der Zande *et al.*, 1984).

CONCLUSIONS

The increased opportunities for woodland creation on better quality land, coupled with advances in tree establishment techniques on disturbed sites, may allow for greater flexibility in design and subsequent management. A better appreciation of the environmental and wildlife benefits of woodland mean that the creation or restoration of 'natural' woodlands is now an important concern. Often this equates only to the visual aspects of woodland design, but these need to be matched by ecological authenticity, avoiding damage or alteration to existing semi-natural communities and making the most of the ecological potential of the site. Allowing natural processes to shape the structural and species diversity of new woodland is vital in this respect; working with nature and not against it makes both environmental and economic sense.

The aim of this book has been to draw on a wealth of experience, from a wide range of backgrounds, in order to synthesise what is already known or borne out of observation and professional judgement. It highlights many issues at the landscape scale of operation, offering practical and sound ecological advice on the integration of woodland with other features, in an approach designed to enhance or maximise biological diversity. The influence of design and subsequent intervention on various taxonomic groups is well understood from research on existing woodland, and can be applied to newly created woodland. Despite this, further refinement is needed, as there remain gaps in our knowledge of the autecology of many species and the functioning of whole communities. The creation of new woodland on a large scale provides ideal opportunities for research and long-term monitoring of successional processes.

REFERENCES

Anderson, M.A. (1989). Opportunities for habitat enhancement in commercial forestry practice. In: Buckley, G.P. (ed). *Biological Habitat Reconstruction.* Belhaven Press, London, 129–146.

Bibby, C.J., Aston, N. and Bellamy, P.E. (1989). Effects of broadleaved trees on birds of upland conifer plantations in North Wales. *Biological Conservation*, **49**, 17–29.

Buckley, G.P. (1994). The effects of edge-management practices on the vegetation of lowland plantations and woods. Unpublished contract report (Project Y29/P4) to the Forestry Authority Research Division, Alice Holt Lodge, Farnham, Surrey.

Buckley, G.P. and Knight, D.G. (1989). The feasibility of woodland reconstruction. In: Buckley, G.P. (ed.). *Biological Habitat Reconstruction.* Belhaven Press, London, 171–188.

Connell, J.H. and Slatyer, R.O. (1977). Mechanisms of succession in natural communities and their role in community stability and organization. *American Naturalist*, **111**, 1119–1144.

Cunningham, M.F.B. (1993). The introduction and establishment of understorey vegetation in woodland: a review. Unpublished contract report (RD914CON.29) to the Forestry Authority Research Division, Alice Holt Lodge, Farnham, Surrey.

Ferris-Kaan, R. and Patterson, G.S. (1992). *Monitoring Vegetation Changes in Conservation Management of Forests.* Forestry Commission Bulletin 108. HMSO, London.

Forestry Authority (1994). *Lowland Landscape Design Guidelines*, 2nd edn. HMSO, London.

Forestry Commission (1990). *Forest Nature Conservation Guidelines.* HMSO, London.

Gill, D.S. and Marks, P.L. (1991). Tree and shrub seedling colonisation of old fields in central New York. *Ecological Monographs*, **61**, 183–205.

Grime, J.P., Hodgson, J.G. and Hunt, R. (1988). *Comparative Plant Ecology.* Unwin Hyman, London.

Grubb, P.J. (1977). The maintenance of species richness in plant communities: the importance of the regeneration niche. *Biological Reviews*, **52**, 107–145.

Helliwell, D.R. (1993). The patterns of nature. *Landscape Design*, **220**, 18–20.

Hibberd, B.G. (ed.) (1988). *Farm Woodland Practice.* Forestry Commission Handbook 3. HMSO, London.

Le Duc, M.G., Sparks, T.H. and Hill, M.O. (1992). Predicting potential colonisers of new woodland plantations. *Aspects of Applied Biology*, **29**, 41–48.

Marsh, S. (1993). *Nature Conservation in Community Forests.* Ecology Handbook 23. The London Ecology Unit, London.

Moffat, A.J. and Bending, N.A.D. (1992). *Physical Site Evaluation for Community Woodland Establishment.* Research Information Note 216. Forestry Commission, Edinburgh.

Myster, R.W. (1993). Tree invasion and establishment in old fields at Hutcheson Memorial Forest. *Botanical Review*, **59**, 251–272.

Peterken, G.F. (1994). New Native Woodlands: an Ecological Analysis. A Commissioned Research Report to The Forestry Authority, Edinburgh.

Peterken, G.F. and Game, M. (1984). Historical factors affecting the number and distribution of vascular plant species in the woodlands of central Lincolnshire. *Journal of Ecology*, **72**, 155–182.

Rodwell, J.S. and Patterson, G.S. (1994). *Creating New Native Woodlands.* Forestry Commission Bulletin 112. HMSO, London.

Soutar, R.G. (1991). Native trees and shrubs for new woodlands in Scotland. *Scottish Forestry*, **45**, 186–194.

Thompson, K. and Grime, J.P. (1979). Seasonal variation in the seed banks of herbaceous species in ten contrasting habitats. *Journal of Ecology*, **67**, 898–921.

Thornber, K.A. (1993). An assessment of the impact of some vertebrate herbivores on regeneration in newly created lowland forest rides. Internal Forestry Authority Research Division Report. The Forestry Authority, Farnham.

van der Zande, A.N., Berkhuizen, J.C., van Latesteijn, H.C., ter Keurs, W.J. and Poppelaars, A.J. (1984). Impact of outdoor recreation on the density of a number of breeding bird species in woods adjacent to urban residential areas. *Biological Conservation*, **30**, 1–39.

Watkins, C. (1990). *Woodland Management and Conservation*. David and Charles, Newton Abbot.

Welch D., Staines, B.W., Catt, D.C. and Scott, D. (1990). Habitat usage by red (*Cervus elaphus*) and roe (*Capreolus capreolus*) deer in a Scottish Sitka spruce plantation. *Journal of Zoology*, **221**, 453–476.

Species Index

Subject Index